Nothing But Performance

The TS-590S

Kenwood has essentially redefined HF performance with the TS-590S compact HF transceiver. The TS-590S RX section sports IMD (intermodulation distortion) characteristics that are on par with those "top of the line" transceivers, not to mention having the best dynamic range in its class when handling unwanted, adjacent off-frequency signals.*

- **HF-50MHz 100W**
- **Digital IF Filters**
- **Built-in Antenna Tuner**
- Advanced DSP from the IF stage forward
- 500Hz and 2.7KHz roofing filters included
- Heavy duty TX section

• 2 Color LCD

KENWOOD

Customer Support: (310) 639-4200
Fax: (310) 537-8235

Scan with your phone to download TS-590S brochure.

www.kenwoodusa.com

ISO9001 Registered
Professional Systems Business Group
JVCKENWOOD Corporation
ADS#13514

* For 1.8/3.5/7/14/21 MHz Amateur bands, when receiving in CW/FSK/SSB modes, down conversion is automatically selected if the final passband is 2.7KHz or less.

A Portable Two-Element 6 Meter Quad Antenna

Assemble or knock down this antenna in less than 5 minutes. Who says Field Day is only about HF?

Pete Rimmel, N8PR

Constructed from PVC tubing, this portable 6 meter antenna disassembles easily for storage, and is easy to assemble at an RV campground or Field Day site. I devised a unique way to lock the spreaders into the center PVC crosses using the antenna wires of the quad. This key feature allows me to tighten the wires so that they won't sag while in use, but lets me loosen them for disassembly and storage in a small space such as in my RV.

Construction

You can construct this quad from parts available at any home improvement store. The PVC dimensions are not critical, and can be within ¼ inch. The center crosses of ¾-inch pipe are compatible with 1-inch Ts that were split to create the boom connections. The inner diameter of the 1-inch T matches the outer diameter of the ¾-inch crosses with a bit of trimming or filing. This larger size PVC is more rigid and makes the antenna more stable.

I used #14 AWG stranded insulated wire for the antenna elements. For me, insulated wire is less troublesome to store in my RV, and won't snag or break while in storage.

Gathering Materials

Cut four 40-inch pieces of schedule 20 × ¾-inch PVC for the driven element spreaders, and four 41¼-inch pieces for the reflector spreaders. From a 10-foot section of schedule 20 × 1 inch PVC, cut two 11-inch pieces for the boom. Use the rest for the vertical mast.

You will also need two ¾-inch PVC crosses (these have four openings); two 1-inch PVC Ts (these have three openings, see Figure 1); one 1-inch PVC cross; and six 1¾-inch diameter stainless steel hose clamps. Cut an 18-foot, 10½-inch length of insulated #14 AWG stranded wire for the driven element and a 19-foot, 5¾-inch length for the reflector. Figure 2

shows the antenna disassembled, with the parts ready for storage.

You'll also need three 2-inch long pieces of ⅜-inch heat shrink tubing, coax sealant, black electrical tape, Velcro® or similar hook-and-loop fastener strips, and a length of 52 Ω coax to reach to your station.

Preparing the Boom

Cut two pieces of the 1-inch PVC pipe to 11 inches long. Place the 11-inch long 1-inch diameter PVC tubes into the 1-inch cross which also connects to the boom, as seen in the lead photo. The rest of the 1-inch PVC goes into the lower part of the T to form the mast.

Make a pair of "five opening fixtures" for connecting the four cross pieces to each end of the boom as follows. Make two "half-Ts" by cutting the long side of

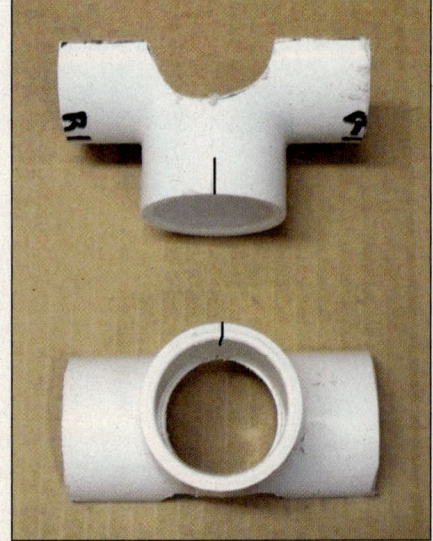

Figure 1 — Split a PVC T and file away material so that with a four-way PVC cross you have a fixture with five openings.

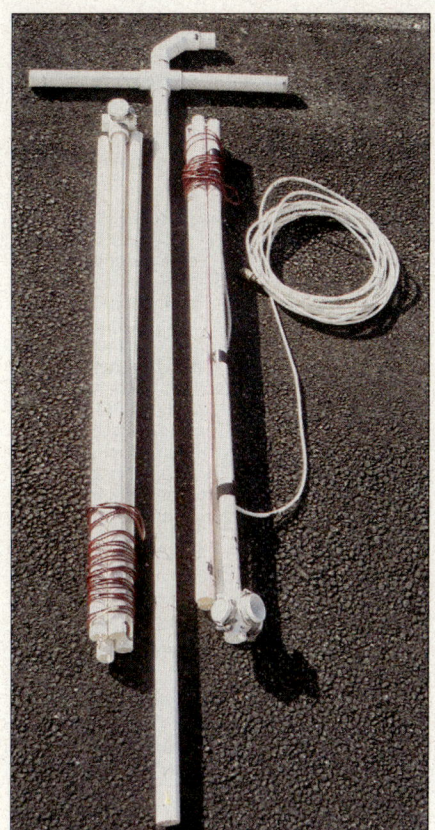

Figure 2 — Easy-to-stow disassembled antenna components.

form an upside down J on both sides of each tube. Be sure to leave PVC in the gap between the holes on the right and the single hole on the left so that the wire can lock in two positions. Carefully "connect the dots" with a drill, and remove all the material between the holes. This forms the unique adjustable J feature of the wire guide on seven spreaders.

Assemble the Reflector
Strip 1 inch of insulation from each end of the longer #14 AWG wire (19 ft 5¾ in) and twist the stranded wire to keep it neat. Thread the wire through the eight J holes of the four 41¼-inch spreaders. Place a 2-inch piece of heat shrink tubing over one end of the wire. Overlap the two ends by 1 inch, twist and solder them together. Slide the heat shrink over the soldered connection and heat it to waterproof the connection. You now have a loop with the wire going through the four outer ends of the spreaders. Position the wire in the inner part of the J on all four spreaders.

Insert the four spreaders into one of the two ¾-inch crosses. After you seat the pieces snugly into the crosses, move the wire to the outer part of the J to tension the wire element. You might not need to tension the wire on all four spreaders. Twist the pipes so that the holes line up and the wire lies in a flat loop when assembled. Your reflector is now completed.

Assemble the Driven Element
Strip 1 inch of insulation from each end of the insulated 18 foot, 10½-inch long #14 AWG wire. Prepare a length of 52 Ω coax (RG8-X, RG-58, or RG 213) by stripping the outer jacket back about 1½ inches. Comb the braid out so that it can be separated from the center conductor and twist the braid into a single strand. Cut it to a 1-inch length. Strip ¾ inch of the insulation away from the center conductor. These dimensions contribute to the overall length of the driven element.

Figure 3 — The five-opening fixture ties the boom end to the four spreaders.

Thread the insulated #14 AWG wire through the holes in the four 40-inch spreaders. Be sure that the bare ends of the wire are near the spreader without the J cut into the end. This is where you connect the coax. Slide a 2-inch piece of shrink tube over each end of the #14 AWG wire. Carefully wrap the center conductor of the coax to one stripped end of wire and solder it. Do not overheat. Wrap the other wire around the braid of the coax and solder it in place. Slide the two pieces of shrink wrap over the soldered connections and heat them. Seal the remaining area of the coax braid with a small amount of coax sealant and cover with black tape to completely waterproof the connection (see Figure 5). Use some loops of Velcro (or black tape) to secure the coax to the PVC tubes. Assemble the driven element by placing the four spreaders into a ¾-inch cross, and place the wires in the J cuts to tension the wire and rotate so that the loop is flat.

Assembling the Antenna in the Field
It is easy to field-assemble and disassemble the antenna; see also the *QST* in Depth web

a 1-inch T with a hacksaw as in Figure 1. Cut a slit in the round part of this fitting. Shape the cut area, and remove a half-circle of material using a rasp or file so that it will fit snugly over the ¾ inch cross. Use two hose clamps to fit each half-T to each PVC cross (see Figure 3). The hose clamps should pull the T snug to the smaller cross without deforming it. A third hose clamp compresses the slit and holds the element on the boom.

Making the Spreaders
Drill 5/64-inch holes ½ inch from the end of each of the eight spreader pieces, going through both walls of the pipe. Take care to center these holes so the wire will go straight through the center of the spreader. Cut J-shaped openings in three of the 40-inch long spreaders and all four of the 41¼-inch spreaders. Following the progression of Figure 4 from left to right, drill a second hole in each spreader at about a 30° rotation and about 1¼ inches in from the end. Again, be sure to drill straight through both walls of the pipe. Draw some guidelines on each of the seven pieces of PVC. Drill a series of holes to

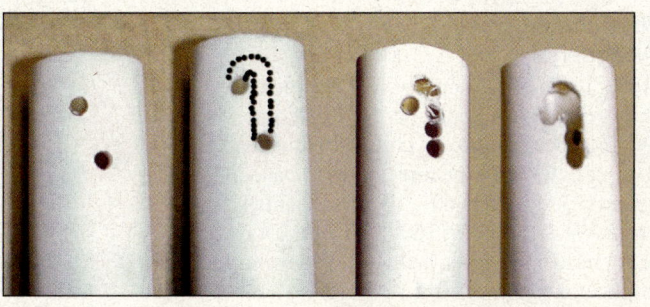

Figure 4 — Steps for making the J holes.

page.[1] Separate the two elements and lay the reflector parts on the ground. Be sure that the wire is in the long part of the J holes. Place the four elements into the cross. Move the wires into the shorter part of the J holes to tension the wire and keep the spreaders from dropping out of the cross. Slide the director onto the boom and partially tighten the hose clamp over the slit in the "half-T" to secure the element to the boom. Assemble the driven element as you did the reflector and mount it onto the boom using another hose clamp. Be sure to orient the elements in the polarization you want before tightening the hose clamps completely. Route the coax along the element and boom to the mast using Velcro loops to keep it in place.

When you take the antenna apart, leave the cross piece attached to one of the four spreaders for storage. Wrap the element wires around the four spreaders to keep them in a neat package.

[1]www.arrl.org/qst-in-depth

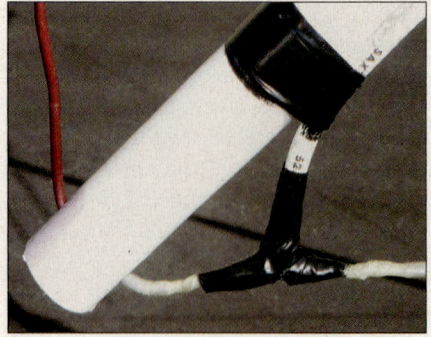

Figure 5 — Detail of the coax feed line connection.

Final Comments

You can easily choose the polarization you want. For vertical polarization, place the driven element so that the coax is on a horizontal spreader, and the antenna is in a "diamond" configuration. For horizontal polarization, place the diamond shape so that the coax is at the bottom of the element. You can also orient the elements for 45° polarization.

This antenna is designed to resonate at 50.1 MHz. The VSWR should be less than 1.5:1 from 50.0 to 50.6 MHz. The element spacing is designed to give a 50 Ω match.

Photos by the author.

Amateur Extra class licensee Pete Rimmel, N8PR, was first licensed in 1960 as KN8UNP in Cleveland, Ohio. He is a Life Member of ARRL and QCWA, holds DXCC #1, 10 Band DXCC, and 5 Band WAZ. Pete is President of Local Chapter #69 of QCWA and served as officer and director of the South Florida DX Association. He earned a BSChE from the University of Cincinnati, and is a retired USCG Master of Passenger Vessels. He recently retired as an NFPA Certified Marine Chemist. Pete enjoys DXing and contesting on 160 to 2 meters, satellite operation, and EME. Other hobbies include sailing, scuba diving, and folk singing. You can reach Pete at n8pr@bellsouth.net.

For updates to this article, see the *QST* Feedback page at www.arrl.org/feedback.

Did you enjoy this article? Cast your vote at www.arrl.org/cover-plaque-poll

New Products

RemoteShack Client Microphone Interface

The RemoteShack RBC-CMI-1 Client Microphone Interface works together with the RemoteShack remote base controller to provide a standard push-to-talk noise canceling microphone to control your cell phone, laptop, or tablet. It will work when making a direct-dialed call to the RemoteShack base or with any smartphone, laptop, or tablet running *Skype* over the Internet. The RBC-CMI-1 also has a speaker output for a larger portable speaker for better quality reception. It comes with a microphone and can interface to most iPhone and

Android devices via the four-pole, 3.5-mm common jack. In addition, the RBC-CMI-1 automatically "keeps alive" the link to the RemoteShack host controller. Price: $249.95. For more information, visit **www.mfjenterprises.com**.

HamCall Adds Call Sign Listings from 1921

Buckmaster's HamCall call sign database now includes all 10,796 US call signs from the year 1921. Included are various pioneers such as 1AF (Harvard Wireless Club), 1AW (H.P. Maxim), and 6BB (University of California Radio Club.) All 1921 data is searchable by name, street, city, state, and by specifying the call sign and year (for example, 1AW:1921). These 1921 archives are in addition to the 1954, 1960, 1969, 1983, 1995, 2000, 2005, 2010, and current data already available on HamCall. Price: HamCall DVD with more than 9.8 million call signs, $50, including 6 months of updates and **HamCall.net** website access, or $80 for 12 months of updates and **HamCall.net** access. For more information, visit **hamcall.net**.

Lightweight Antenna Tuner from SOTAbeams

The Hi-Tee Tuner from SOTAbeams has been optimized for portable operation, but could be used in home stations as well. It uses an air-core coil and includes space on the front to fill in the settings for each band. The tuner weighs 5.3 oz, covers 60 – 10 meters, and is rated for up to 20 W. Price: about $68 (without tax for US/Canada). For more information and ordering details, visit **www.sotabeams.co.uk**.

ARRL's
VHF/UHF Antenna Classics

Practical designs and construction details from the pages of *QST*

Compiled by Steve Ford, WB8IMY

Production: Michelle Bloom, WB1ENT
Paul Lappen

Cover Design: Sue Fagan

Proofreader: Kathy Ford

Published by:
ARRL The national association for *AMATEUR RADIO*
225 Main Street • Newington, CT 06111-1494

Copyright © 2003 by

The American Radio Relay League

Copyright secured under the Pan-American Convention

International Copyright secured

This work is publication No. 295 of the Radio Amateur's Library, published by the League. All rights reserved. No part of this work may be reproduced in any form except by written permission of the publisher. All rights of translation are reserved.

Printed in USA

Quedan reservados todos los derechos

ISBN: 0-87259-907-8

First Edition

Contents

Foreword

50 MHz

Six Meters from Your Easy Chair	Richard Stroud, W9SR	1-1
2 × 3 = 6	L. B. Cebik, W4RNL	1-3
A 6-Meter Quad-Turnstile	L. B. Cebik, W4RNL	1-6
Wire Gain Antennas for 6 Meters	J. Robert Witmer, W3RW	1-11

144 MHz

A Compact Two-Element, 2-Meter Beam	Bob Cerreto, WA1FXT	2-1
The Mini-Five Beam	Lee Aurick, W1SE	2-5
Build Your Own 2-Meter Beam	Dale Botkin, NØXAS	2-7
A "Rope Ladder" 2-Meter Quagi	Jim Ford, N6JF	2-9
Try a "Dopplequad" Beam Antenna for 2 Meters	Keith Kunde, K8KK	2-12
A Five-Element 2-Meter Yagi for $20	Ron Hege, K3PF	2-16
A 2-Meter Phased-Array Antenna	Harold Thomas, K6GWN	2-20
7 dB for 7 Bucks	Nathan Loucks, WBØCMT	2-23
Build a Portable Ground Plane Antenna	Zack Lau, W1VT	2-25
Constructing a Simple 5/8-Wavelength Vertical Antenna for 2 Meters	J. Bauer, W9WQ	2-27
A Glass-Mounted 2-Meter Mobile Antenna	Bill English, N6TIW	2-29
Build a Weatherproof PVC J-Pole Antenna	Dennis Blanchard, K1YPP	2-33
The W3KH Quadrifilar Helix Antenna	Gene Ruperto, W3KH	2-36
Try Copper for 2 Meters–The Cu Loop	Dick Stroud, W9SR	2-40
A Five-Element Quad Antenna for 2 Meters	Jim Reynante, KD6GLF	2-43
Recycling TV Antennas for 2-Meter Use	Ronald Lumachi, WB2CQM	2-46
A Portable Quad for 2 Meters	R.J. Decesari, WA9GDZ	2-49

432 MHz

Optimum Design for 432 MHz–Parts 1 and 2	Steve Powlishen, K1FO	3-1
A 432-Yagi for $9	Dave Guimont, WB6LLO	3-13

902 MHz

A 902-MHz Loop Yagi Antenna	Don Hilliard, WØPW	4-1

Multiband

An LPDA for 2 Meters Plus	L. B. Cebik, W4RNL	5-1
A 146- and 445-MHz J-Pole Antenna	Andrew Griffith, W4ULD	5-6
An Easy, On-Glass Antenna With Multiband Capability	Robin Rumbolt, WA4TEM	5-10
A VHF/UHF 3-Band Mobile Antenna	J.L. Harris, WD4KGD	5-13
The DBJ-1: A VHF/UHF Dual-Band J-Pole	Edison Fong, WB6IQN	5-15
Another Way to Stack VHF/UHF Yagis	Brian Beezley, K6STI	5-18
An Easy Dual-Band VHF/UHF Antenna	Jim Reynante, KD6GLF	5-21

Foreword

Hams never run out of ideas for antennas, especially for frequencies above 50 MHz where designs are smaller and more physically manageable. In *ARRL's VHF/UHF Antenna Classics*, we've gathered a collection of articles published in *QST* magazine between 1980 and 2003. In that 23-year-span, we have published VHF/UHF antenna designs ranging from the simplest ground planes to log-periodic dipole arrays.

The chapters of *ARRL's VHF/UHF Antenna Classics* are grouped by frequency bands from 50 to 902 MHz. This allows you to quickly find the design you need for a particular application. In those few instances where the author provided a design revision (or correction) after the article was published, that revision is *included* in the article as presented in this book.

For RF-safety information and fundamental antenna theory, you'll want to refer to *The ARRL Antenna Book*. See the latest issue of *QST* for other ARRL antenna-related publications, or visit our on-line bookstore at **www.arrl.org/catalog**. Please take a few minutes to give us your comments and suggestions on this book. There's a handy Feedback Form for this purpose at the back, or you can send e-mail to: **pubsfdbk@arrl.org**.

Our thanks to the many authors whose work appears in this book. Without their willingness to share their knowledge with the amateur community, *ARRL's VHF/UHF Antenna Classics* would not exist.

Dave Sumner, K1ZZ
Executive Vice President
Newington, Connecticut
August 2003

Chapter 1
50 MHz

Six Meters From Your Easy Chair	Richard Stroud, W9SR	1-1
2 x 3 = 6	L. B. Cebik, W4RNL	1-3
A 6-Meter Quad-Turnstile	L. B. Cebik, W4RNL	1-6
Wire Gain Antennas for 6 Meters	J. Robert Witmer, W3RW	1-11

By Richard Stroud, W9SR

From *QST*, January 2002

Six Meters from Your Easy Chair

Take one abandoned lawn chair, add some ham ingenuity, and voilà: an effective 6 meter squalo.

If you have a discarded aluminum folding lawn chair, chances are you have the essential elements of an effective 6 meter antenna. After one of our lawn chairs was scrapped because of deteriorating fabric, the aluminum legs were given a rebirth as the elements for a 6 meter squalo. See Figures 1 and 2.

The thin wall tubing measures 0.975 inch OD and the critical 90 degree bends have been neatly done by the chair manufacturer. You will only need to cut the tubing to the dimensions shown. A 12-inch length of 1 inch ID tubing telescopes over the elements to join the two sections. A 3-inch length of Teflon is inserted inside the tubes at the opposite side to stabilize the assembly. One inch Teflon rod (available from Small Parts, Inc[1]) can be turned down as necessary by any machine shop. The entire assembly is bolted together with stainless-steel hardware.

The two capacitance discs are of 0.050 inch aluminum sheet and are cut to a diameter of $3^3/_4$ inches. A center hole is cut in each disc to just clear the tubing OD. Small L brackets hold the fixed disc to the tubing using the same screw that attaches the Teflon spacer. The other disc, which is adjustable, also uses two brackets but one leg of each bracket is extended parallel to the element and clamped in place with a stainless steel hose clamp. The capacitor plates should be parallel and will be about $^{15}/_{16}$ inch apart if the dimensions are followed.

The gamma assembly is made from a $7^1/_4$-inch length of 0.225 in OD thin-wall aluminum hobby tubing with the center element being a $9^3/_8$-inch length of 0.125 inch-diameter (8 gauge) soft copper wire. Teflon sleeving over the wire makes for a good fit inside the aluminum tube. The sleeving should extend $^1/_2$ inch beyond the

[1]Small Parts, Inc, 13980 NW 58th Ct, PO Box 4650, Miami Lakes, FL 33014-0650, tel 800-220-4242 (orders), 305-557-7955 (customer service), **smlparts@smallparts.com**; **www.smallparts.com/**.

Figure 1—**Before:** The humble lawn chair, soon to be transformed.

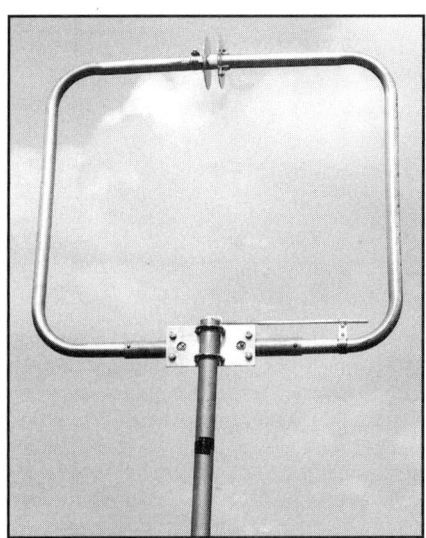

Figure 2—**After:** The 6 meter squalo, ready for action.

Figure 3—Gamma tuning assembly and mounting system. Two $^1/_2$ inch holes are drilled through the aluminum mounting plate to clear the screw heads when the antenna is in the vertical plane.

Figure 4—Detail of the tuning disk construction. The fixed disk is held in place with a screw through the assembly, and the movable disk is clamped using a stainless hose clamp. A Teflon spacer provides support and high voltage insulation.

50 MHz 1-1

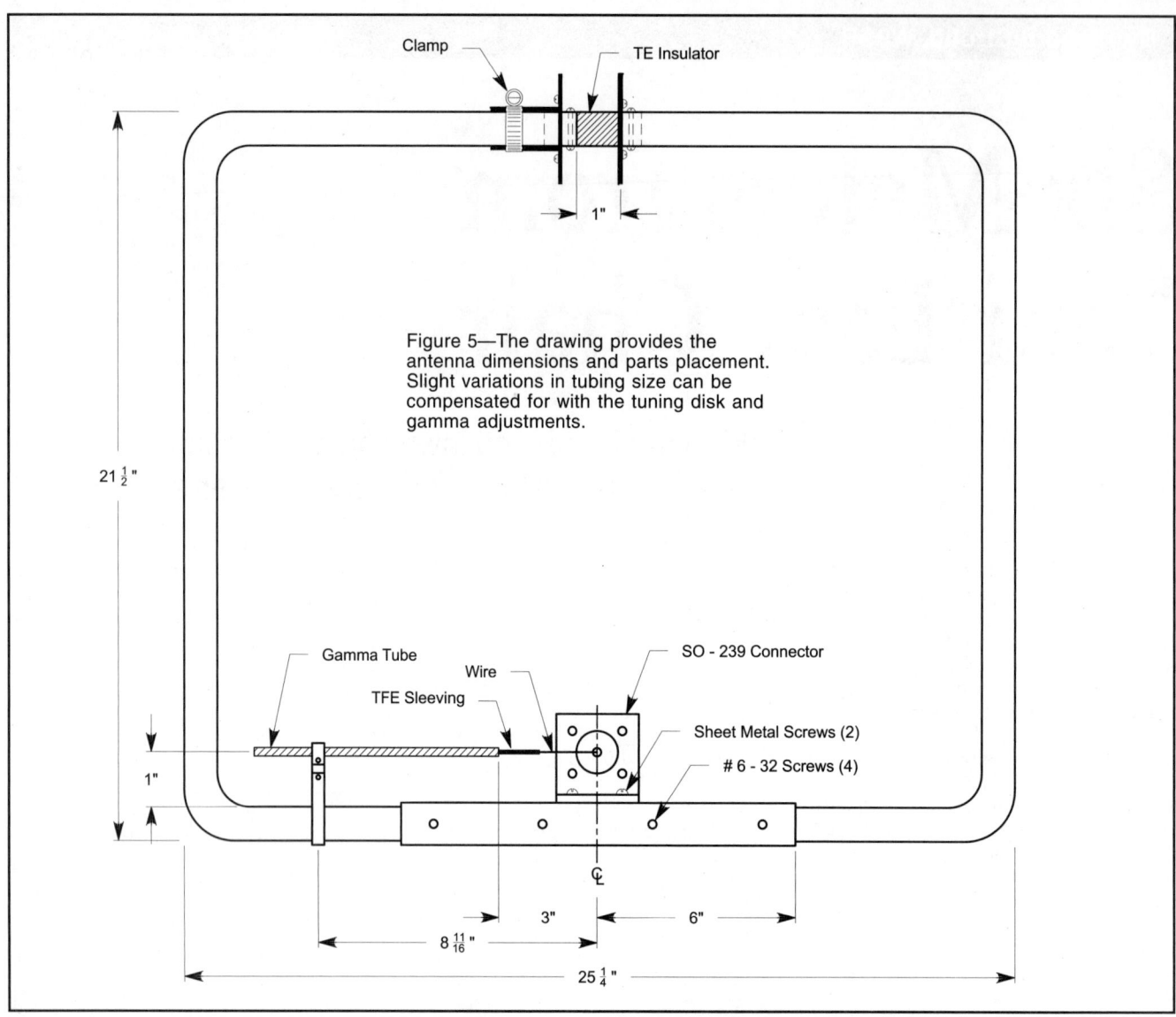

Figure 5—The drawing provides the antenna dimensions and parts placement. Slight variations in tubing size can be compensated for with the tuning disk and gamma adjustments.

length of the wire to provide necessary insulation. A ¹/₂-inch wide double clamp is formed of 0.030-inch aluminum sheet to support the gamma tube 1 inch above the element. The wire is soldered to the center pin of an SO-239 connector mounted in an aluminum bracket at the center of the assembly (see Figure 2). The bracket is fastened to the tubing at the center with sheet metal screws.

Adjustment is made using an MFJ-259B antenna meter with the antenna supported on a mast, in the clear and away from metal objects. Adjust the tuning discs by temporarily loosening the hose clamp and sliding the adjustable disc as necessary. (See Figure 3.) This is a critical adjustment and should be done carefully. If the minimum SWR is above your desired frequency, the discs should be moved closer together and if the frequency is below the desired frequency they should be moved farther apart. Because of the high Q this will probably take several attempts. The hose clamp should be tightened when the frequency is properly set. The next step is to adjust the gamma rod, if necessary, to lower the SWR to 1:1. This is done by moving the gamma attach point and by sliding the tube through the clamp to change the series capacitance. Chances are you will be at a very low SWR with the dimensions shown. After alignment the 2:1 SWR bandwidth is 333 kHz.

The 0.125-inch thick aluminum mounting bracket is 2³/₄ inches by 6 inches and stainless U bolts are used to attach the antenna to a mast. (See Figure 4.) The antenna can be used vertically or horizontally by turning the bracket 90 degrees. The antenna is ideal for mounting horizontal on one leg of a triangular tower, in which case the coax should enter from the top with the connector sealed to prevent moisture entry. Construction details are provided in Figure 5.

The antenna was tested in the vertical plane during the July CQ VHF contest and 47 grids were worked including Cuba and Puerto Rico. This was with the antenna on a test stand at 20 ft above ground and with 40 W output. If placed higher, such as on an existing station tower, the antenna would do an outstanding job. Since it is not directional it is good for monitoring for surprise band openings that could be missed with the station Yagi.

It could be used for mobile operation with a suitable support. Without the mounting hardware, the antenna weighs less than ¹/₂ pound.

By L.B. Cebik, W4RNL
From *QST*, February 2000

2 × 3 = 6

A simple equation? Indeed! Here are two three-element Yagi designs for 6-meter fun!

Yagi antennas provide good forward gain in a favored direction and excellent front-to-back ratio (F/B) for unwanted-signal rejection. A three-element Yagi for 6 meters is a simple construction project and can make use of readily available materials. However, newer antenna builders are often faced with the question, "Which design should I use?"

To help you make the decision, let's look at two quite different designs. Each antenna is a bit over six feet long. One presses for maximum gain and a good F/B, but sacrifices bandwidth. The other achieves total coverage of 6 meters, but surrenders some gain in the process. By comparing the antennas' capabilities with your operating requirements, you can select the one that best suits your needs.

Despite the similar boom lengths, the two designs have quite different profiles, as shown in Figure 1. The wideband model places more distance between the reflector and the driven element and decreases the driven-element-to-director spacing. In contrast, the high-gain model sets the director far ahead of the driven element and decreases the spacing between the driven element and the reflector. The reflector-to-driven-element spacing not only has an impact on gain, but affects the array feedpoint impedance as well. In general, reducing the reflector-to-driven-element spacing lowers the feedpoint impedance.

Gain

Let's first look at the high-gain model to see what we can achieve and what it will cost. A three-element Yagi is capable of exhibiting a free-space gain of 8 dBi with a F/B greater than 20 dB. However, these figures can be sustained for a bandwidth of only little over ±1.5% of the design frequency. Across this span, the antenna's gain tends to increase, while the F/B peaks at over 25 dB near the design frequency.

Our sample high-gain Yagi is adapted from an optimized 20-meter design by Brian Beezley, K6STI. His original design covers all of 20 meters, but that band is narrow compared to 6 meters. When we scale the antenna for 51 MHz, its bandwidth is only about 1.5 MHz while retaining the desired operating characteristics. Table 1 shows the antenna dimensions for a design using ½-inch-diameter tubing. Single-diameter elements are quite practical in VHF Yagis, but before we're finished, we'll see what to do should we decide—or need—to use *two* tubing sizes for each element.

Table 2 shows the antenna's anticipated performance characteristics, as modeled using *NEC 4*. The driven-element length is set near resonance on 51 MHz, and the feedpoint impedance is about 25 Ω. That value isn't a direct match for the 50-Ω coaxial cable normally used in amateur installations. If we shorten the driven element, we can install a beta match. If we lengthen the driven element, we might use a gamma match or a **T** match. If we leave the driven element length as is, we could employ a ¼-λ, 37-Ω matching section

Table 1
Element Lengths and Spacing for the High-Gain 6-Meter Design with ½-inch-Diameter Elements

Element	Length (inches)	Spacing from Reflector (inches)
Reflector	114.26	—
Driven Element	108.96	37.8
Director	102.43	77.94

Table 2
Modeled Performance of the High-Gain 6-Meter Design from 50 to 52 MHz

Frequency (MHz)	Gain (dBi)	F/B (dB)	Feedpoint Impedance (R ± jX ý)	25-Ω SWR
50	7.92	16.55	26.9 – j 20.2	2.14
50.5	8.07	22.59	26.4 – j 11.6	1.57
51	8.24	25.86	24.9 – j 2.4	1.10
51.5	8.43	19.33	22.8 + j 7.8	1.40
52	8.64	14.66	20.3 + j 19.2	2.34

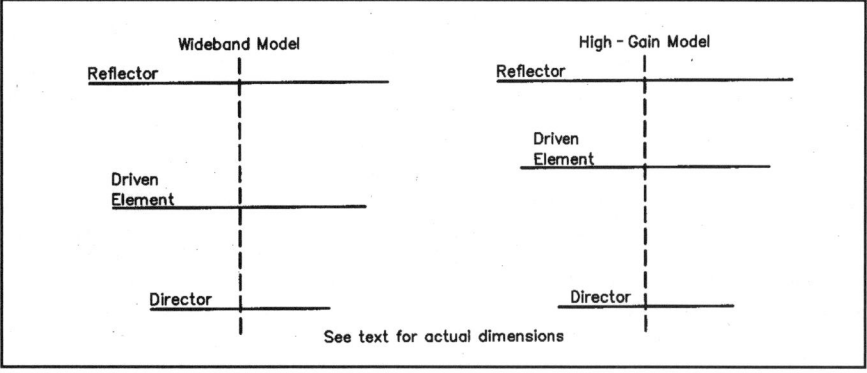

Figure 1—General outline of the wideband and high-gain three-element 6-meter Yagis.

50 MHz 1-3

Table 3
Element Lengths and Spacing for the Wideband 6-Meter Design with ½-inch-Diameter Elements

Element	Length (inches)	Spacing from Reflector (inches)
Reflector	116.80	—
Driven Element	108.10	40.7
Director	96.10	73.5

Table 4
Modeled Performance Figures for the Wideband (50 to 54 MHz) 6-Meter Design

Frequency (MHz)	Gain (dBi)	F/B (dB)	Feedpoint Impedance (R±jX Ω)	50-Ω SWR
50	7.00	14.90	48.4 − j21.2	1.54
51	6.92	18.08	51.9 − j9.9	1.22
52	6.96	20.31	51.9 + j1.7	1.05
53	7.13	21.02	48.8 + j15.0	1.35
54	7.44	18.40	43.0 + j31.1	1.96

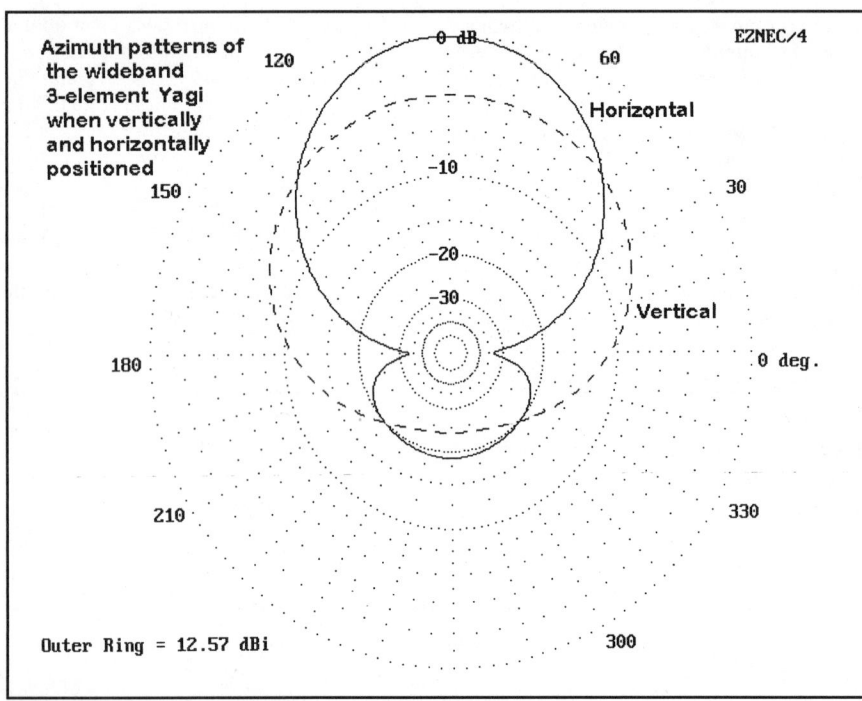

Figure 2—Modeled azimuth patterns for the wideband three-element 6-meter Yagi in horizontal and vertical positions at 30 feet above average earth.

having relatively constant performance all the way. Although this 4-MHz span represents a ±4% bandwidth relative to a design frequency, we can redesign the Yagi to achieve this goal. However, we'll pay for the bandwidth with reduced gain and a lower peak F/B. The gain drops about 1 dB and the F/B is perhaps 5 dB off the peak.

From the same ½-inch-diameter aluminum tubing, we can build a three-element Yagi with a free-space gain of about 7 dBi and a F/B of up to 21 dB. This antenna exhibits a feedpoint impedance that permits direct connection to a 50-Ω coaxial cable (with a suitable choke to attenuate common-mode currents). The design dimensions shown in Table 3 are adapted from a design for another band originally developed by Bill Orr, W6SAI.[2]

The modeled performance parameters appear in Table 4. Notice that the gain curve is not a single rising line, but has a slight dip toward the low end of the band. The F/B peak has been set at the midband frequency because it tends to taper off fairly equally above and below the design frequency. Most notable are the feedpoint impedance and SWR values. If we insulate the driven element from the boom, we can avoid the use of a matching network altogether.

The wideband model is suited to operators who want to cover the entire 6-meter band. However, effective use may require a mechanical scheme that lets you flip the beam from horizontal to vertical. In the vertical position, as shown in Figure 2, at a height of 30 feet above average ground, the pattern is wider and less strong than when the antenna is used horizontally. Still, these beams are both simple and inexpensive. Hence, you might want to build a high-gain model for the low end of 6 meters and a wideband model to cover the upper 3 MHz of the band.

Figure 3 overlays free-space azimuth patterns of both beams at their design frequencies. The patterns will give you a good idea of their relative performance potentials.

Stepped-Diameter Tubing

The beam dimensions for both models used uniform ½-inch-diameter elements. A common building practice is to use at least two tubing sizes in 6-meter beams. Most often, we start with ½-inch tubing at the center and use ⅜-inch tubing for the element ends. Let's suppose we make the center portions of each element from 6-foot lengths of ½-inch tubing—3 feet of tubing on each side of the boom. What happens to the overall element lengths?

Table 5 compares the element lengths from the boom outward for each beam (commonly called "element half-lengths"). One model uses ½-inch-diameter tubing throughout, and the other uses ⅜-inch-diameter tubing for the ends. The stepped-diameter lengths are chosen so that the antenna performance is essentially the same as with uniform-diameter elements. Note that the

made by connecting two lengths of RG-59 (or RG-11 for high-power operation) in parallel. All of these matching systems are described in *The ARRL Antenna Book*.[1]

The table of projected gain and F/B values shows the rise in gain across the passband, as well as the peak F/B near the design frequency. Notice that the F/B drops rapidly as we approach frequencies only 1 MHz from the design center. For point-to-point communications at the low end of 6 meters, however, the narrow passband—combined with the higher gain—may be just what we need.

The target center frequency can be adjusted up or down within 6 meters by adjusting all three element lengths by the percentage of frequency change. To change the design frequency to 50.5 MHz to cover the 50 to 51-MHz range, increase all lengths by about 1%. If we stay at the low end of the band, we need not change the element spacing or diameter.

Builders who are more interested in raw gain than F/B can scale the performance at 52 MHz (or a bit above) down to the desired frequency. Simply scale the antenna dimensions, as given for the 51-MHz design frequency, to about 50 MHz or just a bit lower. You can adjust the driven-element length to resonance or use your favorite matching system. Changing the driven-element length to vary the feedpoint impedance by as much as 25-30 Ω has very little effect on the other performance figures.

Bandwidth

Suppose we want to cover the entire 6-meter band with a well-matched Yagi

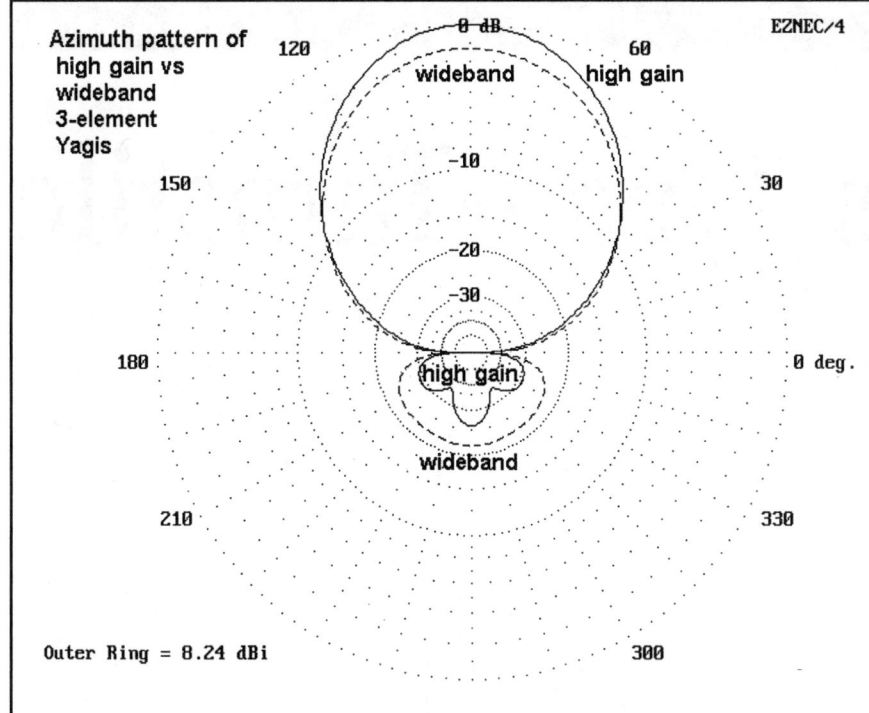

Figure 3—Overlaid models of free-space azimuth patterns for the high-gain and wideband 6-meter designs at their design frequencies.

Table 5
Half-Element Lengths for Uniform Half-Inch and Stepped ½-Inch to ⅜-Inch-Diameter Elements

High-Gain Design

Element	Uniform Diameter Length	Stepped-Diameter Length
Reflector	114.36	116.4
Driven-element	108.96	111.0
Director	102.43	104.0

Wideband Design

Element	Uniform-Diameter Length	Stepped-Diameter Length
Reflector	116.80	118.6
Driven-element	108.10	110.0
Director	96.10	97.6

Note: All dimensions are in inches.
For the stepped-diameter elements, the inner 36-inch length uses ½-inch-diameter tubing, with ⅜-inch-diameter tubing used for the remainder of the element.

element lengths become significantly longer when we step the element diameter downward on the way to the element end. The amount of change differs for each element.

You can calculate the end lengths by subtracting 36 inches from the overall element length. However, be sure to add about three inches per end section to allow for telescoping the tubing.

I'll leave the remaining construction details up to you, since there are many acceptable ways to construct either of these Yagis. Again, *The ARRL Antenna Book* and articles in *QST* and recent editions of *The ARRL Antenna Compendium*[3] are full of good ideas. Simply select those that best fit your available materials and individual skills.

Both of these Yagis—adapted from the work of veteran antenna designers—are good designs. Which you choose will depend on what you want to do on 6 meters during the present sunspot cycle and beyond.

Notes

[1] ARRL Publications are available from your local ARRL dealer or directly from ARRL. Mail orders to Publication Sales Dept, ARRL, 225 Main St, Newington, CT 06111-1494; tel (toll free) 888-277-5289, 860-594-0355; fax 860-594-0303; e-mail to **pubsales@arrl.org**. Check out the full ARRL publications line on the World Wide Web at **www.arrl.org/** and the Bookcase in each issue of *QST*.

[2] Bill Orr, W6SAI, Ham Radio Techniques, "May Perambulation," *ham radio magazine*, May 1990 pp 56-61.

[3] See Note 1.

By L. B. Cebik, W4RNL
From *QST*, May 2002

A 6-Meter Quad-Turnstile

Looking for improved omnidirectional, horizontally polarized performance? This 6-meter turnstile uses the quad loop as a foundation.

Turnstile Principles and Limitations

Figure 1 shows the classic turnstile configuration: Two dipoles at right angles to each other. The main feed line feeds one of the dipoles. A ¼ λ phase line runs from the first dipole and feeds the second. If the phase line is exactly ¼ λ long (or 90° electrically) and if the impedance of that line is a match for the resonant impedance of the individual dipoles (70 Ω), then the second dipole will have a current magnitude that is identical to that of the first dipole but the dipole currents will be 90° out of phase. Proper current phasing is required to obtain a nearly circular pattern with no more than about 1 dB gain variation. This ideal condition is called quadrature.

The standard dipole-turnstile has some limitations. The system feed point impedance is 35 Ω and requires a matching system if the 1.45:1 SWR level is not satisfactory. Special impedance-based systems that simultaneously obtain proper phasing and a 50-Ω match tend to show distorted patterns because, while they present a good match to the main feed line, they fail to achieve the proper current conditions on the two dipoles. Indeed, the SWR curve for any turnstile is so broad that it is useless as an indicator of proper antenna operation.

The dipole-turnstile has a second limitation: If we operate the antenna too far from the design frequency or if we carelessly construct the antenna or phase line, then the pattern will no longer be omnidirectional. Instead, it becomes a bi-directional oval with an increasing differential between maximum and minimum gain as we drift from the design frequency. Unfortunately, SWR will give us no clue to the drift. Careful design and construction are the keys to effective turnstile operation.

A third limitation of the dipole-turnstile results from the fact that the strongest

An overall view of the quad-turnstile on the assembly and test stand. I took this photo during initial tests before adding the perimeter cord and taping down the phase line and feed line.

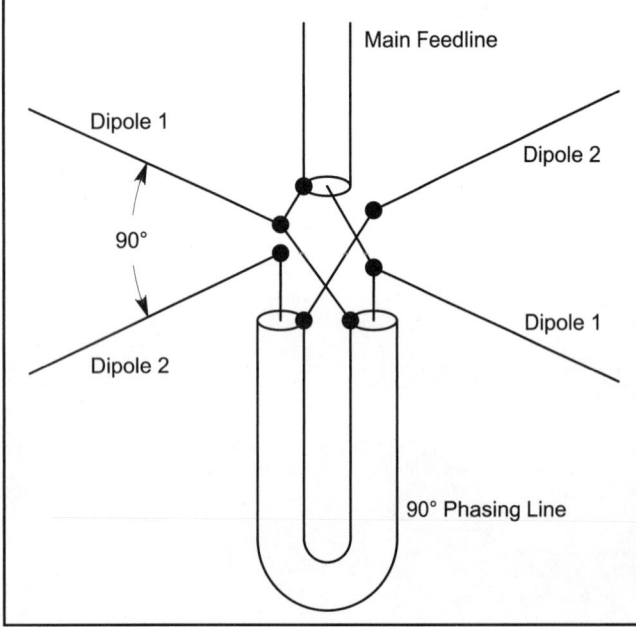

Figure 1—The general outline of a dipole-turnstile. The feedpoint connection scheme also applies to the quad-turnstile described in the text.

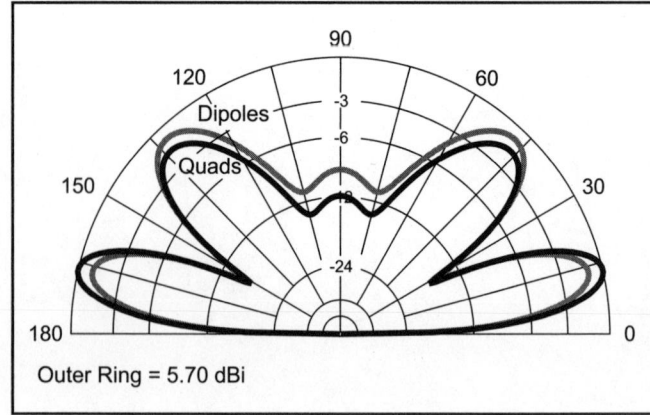

Figure 2—Comparative elevation patterns for a dipole-turnstile and a quad-turnstile when both antennas are 1 λ above ground.

radiation is broadside to the pair of dipoles. In free space, the H-plane radiation is stronger than the E-plane radiation. Ground reflections permit the dipole-turnstile to develop a usable pattern. Figure 2 shows the elevation pattern for the dipole-turnstile. At 1 λ above ground the second elevation lobe is actually stronger than the lowest, the lobe we use for point-to-point 6-meter communications. One wavelength is about 20 feet at 6 meters—a typical height for casual operation.

Figure 2 also shows the elevation pattern for a pair of quad loops in a turnstile configuration that partially corrects the radiation pattern problems. This antenna's lower lobe is the strongest and in addition, shows about 1 dB additional gain. Figure 3 compares the azimuth patterns for the two antennas at an elevation angle of 14°. The quad-turnstile not only has more gain, but its pattern is slightly more circular than that of the dipole-turnstile.

For contrast, Figure 4 shows the azimuth patterns of the lower lobes with the antenna operated 5% off-frequency (53.025 MHz). The effect—a radically distorted pattern—is equivalent to cutting the loops 5% too long. The current magnitudes on the elements of both types of antennas are no longer the same, and the current phase angle between elements is no longer 90°. The quad-turnstile may offer a bit more gain, but it is just as susceptible as the dipole-turnstile to both construction and operating errors.

Regardless of important design sensitivities of turnstiles in general, the quad-turnstile has enough advantages over the dipole-turnstile to warrant consideration if you are just beginning to look for an omni-directional horizontal antenna for 6 meters.

Building a Quad-Turnstile

The photographs will give some perspective views of the basic quad-turnstile, while the sketches of construction details provide dimensions and details. A quad-turnstile will maintain its omnidirectional pattern with about a 1-2% error either in construction or operating frequency. Therefore, the quad-turnstile shown here has been optimized for 50.5 MHz using AWG 14 bare copper wire. Do not use the listed dimensions with insulated wire, since insulation gives antenna wire a velocity factor that ranges from about 0.95 to 0.99, depending upon the thickness and composition of the insulating material. If you choose to use insulated wire, first construct a single quad loop element and bring it to resonance at the design frequency. The dimensions you obtain for this antenna can be copied to the second loop.

Figure 5 shows the dimensions of a quad-turnstile based on a diamond configuration. There is no significant radiation difference between a square quad loop and a diamond-shaped loop. The diamond allowed a very simple form of construction and became the basis for the quad-turnstile. Table 1 is a bill of materials for the antenna.

The Phasing line

A resonant quad loop has a feedpoint impedance of just about 125 Ω. We connect the two loops of the quad-turnstile exactly as we would connect a dipole-turnstile. We need a 90° (¼ λ) phasing line having this characteristic impedance (Z_0). RG-63, carried by the Wireman, is ideal for the job. The resulting antenna feedpoint impedance is 62 Ω, which requires no special matching to a standard 50-Ω main feed line if a 1.25:1 SWR is acceptable. (In practice, the minimum SWR may be about 1.3:1 to 1.4:1, depending upon the reactance introduced by the leads between the loops and the coax connectors.)

A wavelength at 50.5 MHz is 233.72 inches, so a quarter-wavelength is 58.43 inches. However, the listed velocity factor of RG-63 is 0.84. Therefore, 49.1 inches (58.43 × 0.84) will be the correct physical cable length to give a 90° electrical length. When cutting the cable, include the coax connectors in the measurement of total cable length, since they are part of the shielded cable section.

Having prepared the phasing line for the quad-turnstile, let's build the antenna itself. My support structure consists of a center pole and two cross arms. The center pole is 1 inch nominal Schedule 40 PVC. In some areas of the country, white PVC is not adequately protected from ultra-violet degradation, so the gray electrical conduit of the same size may be a better choice. The cross arms consist of ½ inch nominal CPVC, which is strong enough to support the wires with minimal weight. The cross arms pass through holes (at right angles to each other) in the center pole. A 1½-inch #10 stainless steel bolt, lock washer and nut secure each cross arm. The cross-arm holes in the main pole should be immediately above and below one another without overlapping.

At each end of the cross arms I drilled a

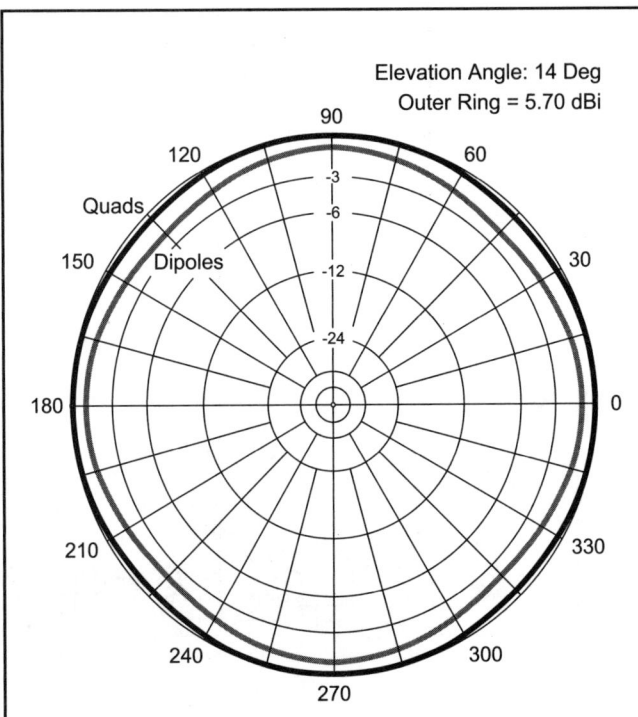

Figure 3—Comparative azimuth patterns for a dipole-turnstile and a quad-turnstile when both antennas are 1 λ above ground. The elevation angle is 14°.

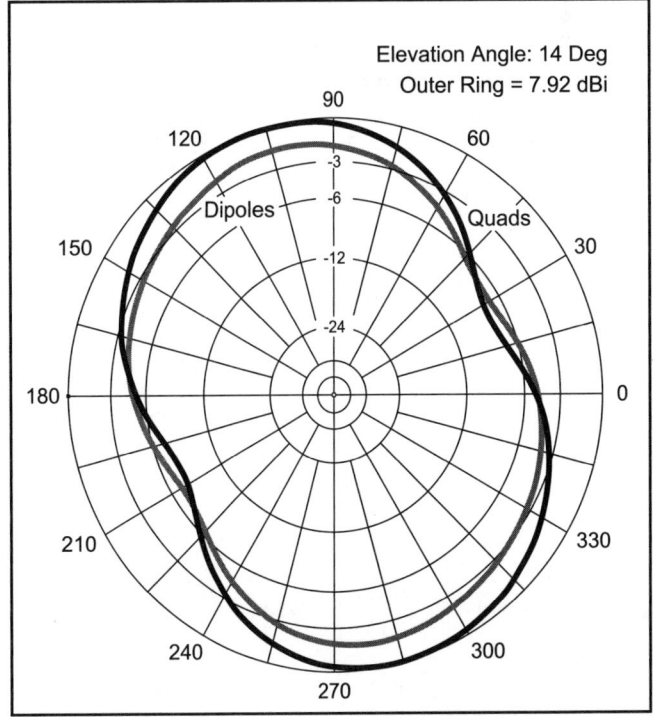

Figure 4—Comparative azimuth patterns for a dipole-turnstile and a quad-turnstile when both antennas are 1 λ above ground at 5% above the design frequency. The elevation angle is 14°.

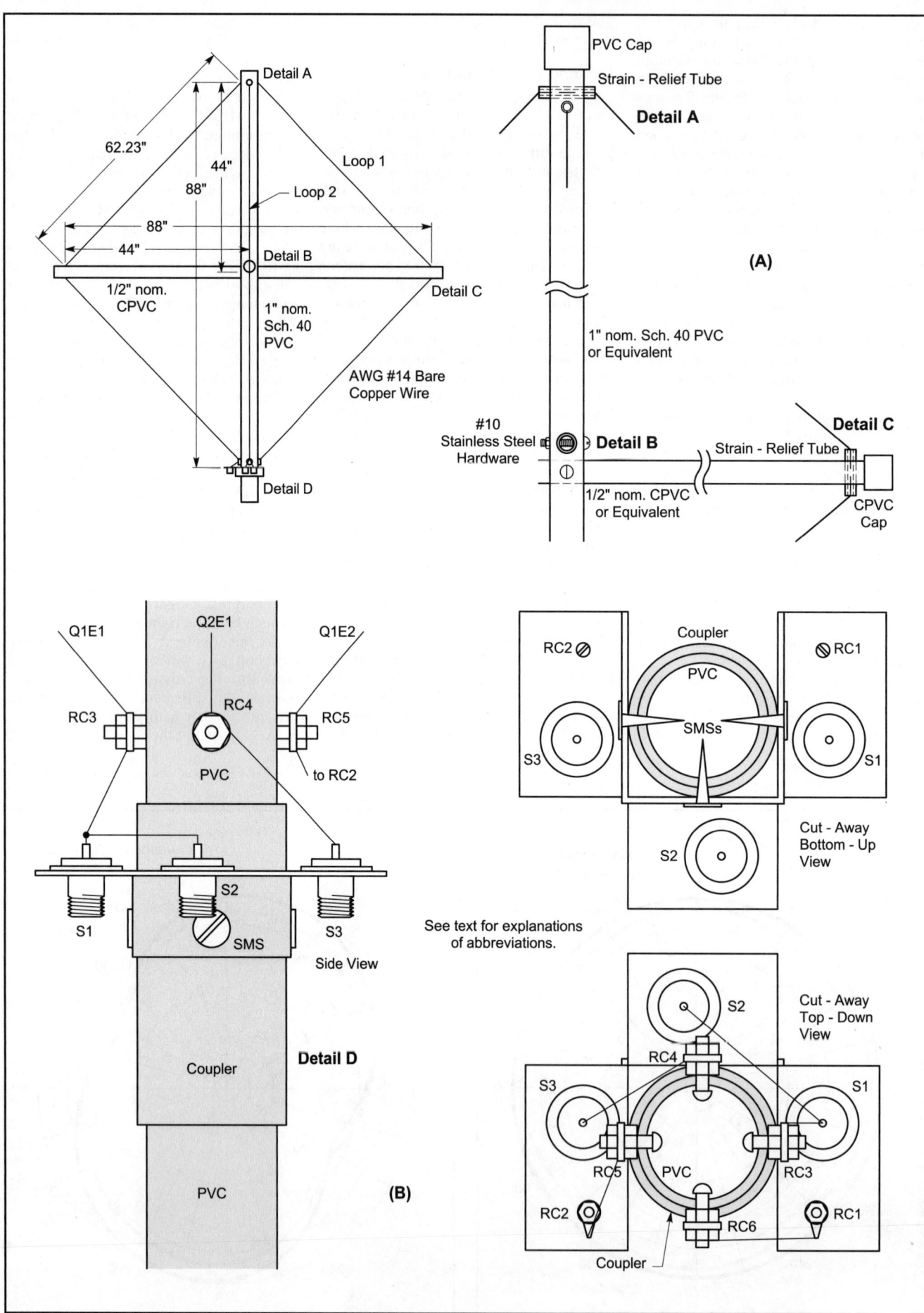

Figure 5—The general outline and dimensions of each quad loop in the quad-turnstile. Details A, B and C show the main post-to-crossarm connection and the slip tubes and end caps at the quad loop corners. Detail D shows several views of the antenna base connections and the coax connector assembly.

A close-up view of the base and connector assemblies. The photo preceded the final waterproofing of all exposed connections.

Table 1
Bill of Materials

Quantity	Material Description
43' min	AWG 14 stranded or solid bare copper wire.
5'	RG-63/U coaxial cable.
2 – 8' sections	1/2" nominal CPVC pipe (gray electrical conduit may be substituted).
1 – 8' section	1" nominal Schedule 40 PVC pipe (gray electrical conduit may be substituted).
4	1/2" CPVC pipe cap.
1	1" PVC pipe cap.
1	1" PVC pipe coupler.
3 to 4"	1" aluminum L-stock (also called angle stock).
3	SO-239 single-hole connectors.
2	#10 1 1/2" stainless steel bolt.
6	#10 3/4" stainless steel bolt.
2	#10 stainless split-O steel lockwasher.
10	#10 stainless steel nut.
8	#10 stainless steel flat washer.
3	#10 stainless steel sheet metal screws.
6 – 2" pieces	Nylon or plastic tubing (see text).
2	#10 ring terminals.

3/16 inch hole. I epoxied a piece of plastic tubing in the hole to reduce abrasion on the antenna wires. Nylon or other plastics, as well as AWG 12 house wiring insulation, will do the job if the #14 wire passes through smoothly. [US Plastic Corp (tel 1-800-809-4214; www.usplastic.com) sells flame retardant and UV-resistant polyethylene tubing, recommended for this application.—Ed.] For ease of handling, I used tubing pieces that are longer than needed. Once the epoxy had set, I trimmed the tube ends about 1/4 inch beyond the cross arm. Figure 5A shows the rough details of the cross-arm and main pole end treatments.

The main support pole has two 3/16 inch wire holes near the top, very close to each other and at right angles. These holes align with the cross arms and pass each quad loop while maintaining a small separation. Both holes receive the strain-relief tubing treatment. The cross arms and the main pole top end have about an extra inch of PVC, permitting the addition of caps to keep water out of the support pipes. For true water-tightness, seal the opening in the main support where the cross arms pass through, as well.

The Loop Wires and Connector Assembly

We can prepare the top three corners of the quad loops simply by measuring, drilling, cementing, and bolting. However, let's proceed more slowly with the rest of the antenna. The first step is to measure the wire for each loop. For AWG 14 bare wire (stranded or solid), the loop circumferences are 20.74 feet or 249 inches as close as your tape measure will permit. Leave (and mark) about 2 inches of extra wire on each end for connection leads. Tin—that is, coat with solder—the first and last 3 inches of each wire, whether stranded or solid.

Set the wire aside and examine Figure 5B. In the following notes, the abbreviations used on the drawing appear in parentheses. At the base of the main PVC post a coupler—a short PVC connector designed to join two sections of PVC pipe—will be installed. We shall also construct a bracket to hold the three coax connectors we need—one for each end of the phase line and one for the main feed line.

The bracket was made from a piece of 1/16 inch thick aluminum L-stock that was 1 inch on a side. I measured the PVC coupler outside diameter and marked the L-stock accordingly. I then drilled 5/8 inch holes in the L-stock top plate for single-hole SO-239 coax connectors (S1, S2, S3). As shown in the drawing, the holes are as close to the open edge of the L-stock as the connector mounting hardware would permit. I also offset them from the center of each section of stock. The offset permits easy installation of the bracket mounting screws, and the near-edge position provides clearance for the male coax connector shell to fit between the SO-239 and bracket. #10 stainless steel hardware provides mounting points for the ring terminals (RC1 and RC2) that will terminate each quad loop on the coax braid side.

Once the drilling is complete, hacksaw through the top plate of the L-stock so that you can bend it around the PVC coupler, making the three-sided bracket. On each side, drill a hole in the L-stock to pass a #10 stainless steel sheet metal screw (SMS). Fit the bracket very close to the top of the coupler and drill smaller holes so that the screws will bite and tap the coupler. Attach the bracket to the coupler with the sheet metal screws. However, do not yet glue the coupler and connector assembly to the main post.

Take one of the loop wires and temporarily clamp or tape it to the post just above the point where you think that you will mount the base assembly. Now thread the wire through the cross arms and main post holes so that both ends approach the base of the main post at the marks you made on the wire indicating the ends of the loop. Mark the post for the four #10 stainless steel bolts and nuts (RC3, RC4, RC5, RC6). From the bolt position, leave enough extra PVC (about 1 1/8 inches) to allow mating with the coupler and then cut off the main post. Drill four holes in the main post and install #10 bolts with a single nut on each one.

Remove the connector bracket from the coupler. Cement the coupler to the main post using good quality PVC cement, maintaining good alignment between the connectors and the cross arm assemblies. Once the PVC cement dries (less than 1 minute), drill through the holes in the coupler into the main post. Reattach the connector bracket. The double thickness of PVC will provide a secure assembly with only three #10 sheet metal screws.

Installing the Loops

Start the installation of the first loop. Leaving enough wire to reach the RC1 or RC2 terminal, wrap a single tight turn of the antenna wire around the connector bolt. This relieves strain from the coax and ground connectors on the base assembly. Add a washer and nut and tighten this connection. As an alternative, you may cut the loop wire and use ring connectors on this bolt. Now add (crimping and soldering) the ring connector that attaches to the connector assembly plate, trimming any excess from the wire-end lead.

At the other end of the wire, similarly attach the loop wire to the proper bolt on the main post. Tension the wire enough so that the loop does not flop about, but not so much as to bend the cross arms. Run the loose wire end lead to the appropriate SO-239 terminal (S1 or S3), trim any excess, and solder. Repeat this process for the second loop. Be sure to add a bridge wire from the main feed line connector (S1) to the first phase line connector (S2).

You can optionally add at each cross arm a thin wire soldered to the loop wire above and below the slip tube and run on the outside of the arm. These wires will tend to keep the cross arms from drooping with time, sun, and wind. For additional resistance to wind effects, you may add a perimeter UV-resistant cord, looping and knotting the cord at the ends of each cross arm. The result is an octahedron with considerable sturdiness, despite the light materials.

Completing the Antenna

When you have finished the basic antenna construction, you should have some leftover 1 inch PVC. Cement one end of this tubing into the bottom of the coupler. To mount the antenna on a standard TV mast, you can cement or bolt a length of this tubing into 1 1/4 inch nominal Schedule 40 PVC to make a fitting that will slip over the mast. A few wraps of electrical tape around the TV mast will provide a tight fit for the larger PVC, while the 1 inch stock will come to rest on the mast top.

Since coaxial cable tends to be heavy, when you install the main feed line and the phase line, tape them to the main mast. A pair of tape wraps about 1 foot apart will provide adequate strain relief for the cables, antenna, and fittings. For minimum interactions among lines, place the phase line on one side of the mast and run the main feed line down the other side. Once you have tested the antenna, waterproof all coax fittings and connections.

Conclusion and the Next Step

I have relied on construction details that will assure an antenna that is close to optimal, because normal ham equipment does not permit very meaningful performance measurements. If you choose to use a different wire or construction scheme, the first step will be to resonate a single quad loop at the design frequency. A single loop will show a 125-Ω resistive impedance with little or no reactance at resonance. You can then use this wire length for both of the quad-turnstile loops. The required wire lengths for bare AWG 18 through AWG 10 are within the recommended 1% tolerance, so you may use the listed dimensions with any of these wire sizes.

The performance expectations that I have shown are based on placing the base of the antenna at a height of 1 λ above ground. Additional height will lower the elevation angle of the lowest lobe and improve performance. There will be a limit to wind survival, however, with the fairly light construction shown.

Beyond the quad-turnstile, for added omnidirectional horizontally polarized gain, we might turn to a pair of dipole-turnstiles vertically stacked about 1/2 λ apart. However, the design of such a stack involves paying close attention to the mutual coupling between the dipoles. That interaction results in the need for a non-standard phaseline Z_0, along with other complexities of in-phase feeding of the arrays. The assembly is also considerably wider and taller than our simple quad-turnstile.

As an intermediate step in the process of improving omnidirectional antenna performance, the quad is a notch better than the dipole. However, the care with which we constructed the quad-turnstile, if applied to a dipole-turnstile, can go a long way toward getting the most out of the more elementary antenna. Phase-feeding elements, even in simple-looking arrays like the dipole- and quad-turnstiles, requires all the care that we can bring to the shop.

By J. Robert Witmer, W3RW
From *QST*, February 1997

Wire Gain Antennas for 6 Meters

Get some *gain* on 6 meters—without investing in a beam and rotator!

In August 1995 I came across an old Clegg Venus 6 meter SSB/CW transceiver (a 1960s vacuum-tube rig). After the radio had sat on my workbench for several months, I finally got around to fixing the previous owner's "design improvements." Soon thereafter, the Venus was on the air!

I had a great time in the 1996 January VHF Sweepstakes and enjoyed the sporadic-E season in the spring and summer of 1996. Until recently, however, I had been using a vertical antenna cut for the FM portion of 6 meters. It was terrible for local SSB work—most SSB and CW operators use horizontal antenna polarization on the VHF bands. During normal groundwave operation you are at a big disadvantage if you operate with the opposite polarization. During long-distance band openings, it doesn't matter quite as much, but the vertical still seemed to be lacking in performance.

Why Not a Beam?

I live in a neighborhood where RFI/TVI/BMI (baby monitor interference) reports can begin just with the *installation* of a new antenna, let alone an actual transmission. Because of the desire to maintain a low antenna profile—plus my unwillingness to make the investment in a tower, beam and rotator—I decided to investigate other approaches to reasonable antenna performance for 6 meter SSB/CW operation.

What follows is the highlights of what I've found. I haven't had the opportunity to thoroughly check all of the antenna possibilities I will describe, but I've included the references and important information so *you* can try these antennas for yourself.

By the way, going to horizontal polarization has made a big difference in BMI so far—baby monitors use vertically polarized antennas!

The Long-Wire Antenna

We don't usually think of long-wire antennas for VHF applications, but they can be used on 6 meters almost as easily as on the HF bands. In fact, a wire is typically not considered "long" until it is several wavelengths long. At 6 meters a wire four wavelengths long is only about 75 feet—a length that will fit in many locations. According to *The ARRL Antenna Book*, an antenna four wavelengths long can have a gain over a dipole of approximately 3 dB (3 dBd) in some directions. The antenna can be fed at the end, or at a *current loop*. Because of matching considerations (I don't have a 6 meter antenna tuner) I chose to use the current-loop approach (see Figure 1). You could make the antenna longer and pick up more gain if you like. An antenna six wavelengths long should have a gain of almost 5 dB, and an antenna 10 wavelengths long should have a gain of approximately 7.5 dB.

Along with an increase in gain, there will be a change in the radiation pattern. You're familiar with the doughnut-shaped radiation pattern surrounding a half-wavelength dipole—that pattern breaks up into a multilobed pattern as the length of the antenna is increased. The bottom line is that you may end up with 3 dB gain in *some* directions with that four-wavelength long wire, but there will also be nulls (where the gain becomes *less* than that of a dipole) in other directions. With a fixed long-wire antenna, you take "potluck" on what your gain will be in the direction of a station you hear—but if you hear him, you stand a fair chance of working him.

Building a Long Wire for 6 Meters

I used the formula 3924/f (where f = frequency in MHz) to determine the overall length of my four-wavelength long wire. That antenna would fit in my 80-foot space. I then determined the current-node point by using the formula 234/f, and feeding the antenna that distance from one end. The radiation impedance at that point is about 130 Ω. The resulting SWR, using a 4:1 balun and 50-Ω cable, should be less than 2:1.

When you cut your wires, always make

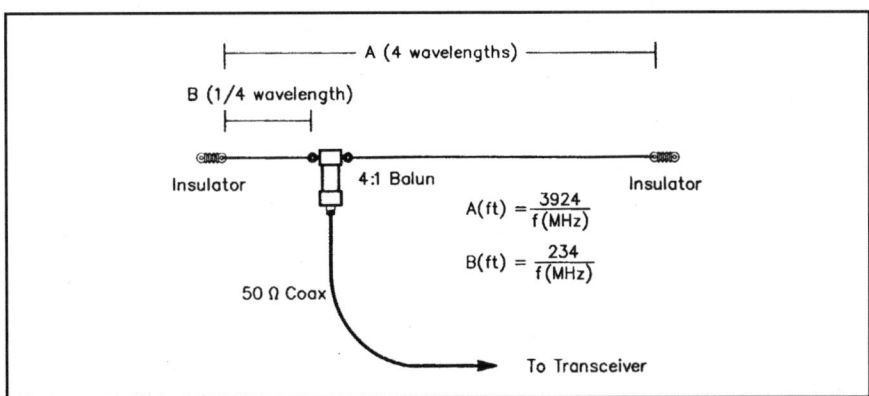

Figure 1—An off-center fed long wire antenna for 6 meters. It's basically just two pieces of wire linked by a 4:1 balun. Choose your antenna's "center frequency" (f) and cut length A using the formula shown. Mark the current node point—¼ wavelength (B) from one end—and cut again. Cut the lengths of both sections a little longer than your calculations call for, so you have a little surplus for adjustment purposes.

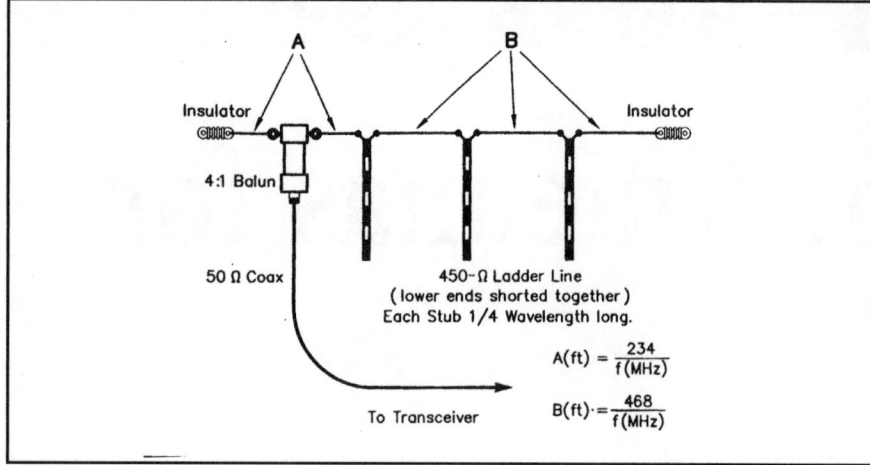

Figure 2—The multielement collinear array uses stubs to get the RF currents in phase. The stubs in this example are made of ¼-wavelength sections of 450-Ω ladder line (shorted at the ends). To calculate the lengths of the stubs, use the formula (246/f)×V, where f is the center frequency of the antenna and V equals the velocity factor of the ladder line you've chosen. The 4:1 balun is once again attached at the current node point.

Multielement Collinear Wire Array

As I mentioned earlier, a four-wavelength wire has gain over a dipole of about 3 dB. The gain isn't higher because the antenna currents are not in phase, creating some field cancellation. To increase the antenna gain, it is necessary to get the RF currents in phase. Figure 2 shows how this can be done using a *collinear wire array*.

The collinear is fed the same way as the long-wire antenna but uses ¼ wavelength shorted *phasing stubs* between the half-wave-length elements. You can make phasing stubs out of common 450-Ω ladder line, but you have to watch the *velocity factor*. Don't let this term scare you. It merely refers to how fast a radio signal travels through a given piece of feed line, expressed as a decimal percentage of the speed the radio signal travels in free space (ie, the speed of light). Velocity factor is an important consideration when you're playing with the phases of radio waves. Depending on its velocity factor, the cable you cut to a mechanical ½ wavelength might be something quite different in an electrical sense.

I've discovered that velocity factors vary between manufacturers of 450-Ω line. Several samples I've tried had velocity factors varying between 0.85 and 0.9—neither of them close to the commonly quoted value of 0.95. Try to verify the velocity factor of the line you plan to use. If you can get your hands on a grid-dip meter, you can couple it to the shorted stub and check the resonant frequency.

Using the phasing-stub technique, I put up four ½ wavelength sections with their accompanying matching sections in the same space my four-wavelength wire occupied. Maximum radiation is off the sides of the antenna. It's difficult to achieve optimum gain because you're bound

them a little longer than the formulas indicate. When you attach the wires to the insulators as shown in Figure 1, wrap the surplus length back on the wires. That way, if you need to lengthen the antenna, you can simply unwrap the extra length.

At the current node you can attach both wires to a commercially made 4:1 balun. Just make sure it is rated for use on 6 meters. (Two examples: The W2FMI-4:1-HBM200 made by Amidon Associates, PO Box 25867, Santa Ana, CA 92799; tel 714-850-4660; fax 714-850-1163. The Centaur baluns sold by Amateur Electronic Supply; tel 800-558-0411.) These baluns tend to be bulky and their weight might make your wire sag unacceptably. If this is the case, attach a run of 300-Ω ladder line at the current-node point. The line should be ½ wavelength at the frequency f. Snake the ladder line back to the 4:1 balun and go from there.

Long-Wire Antenna Performance

I've compared the performance of my 6 meter long wire to my ⅝ wavelength vertical during several band openings. The longwire antenna often performed better! The most noticeable change occurred when I used the long wire for local communication. The difference was substantial. On some weaker signals switching to the vertical would make the signals disappear! My biggest thrill was working the only "double-hop" station I heard during the June VHF contest—and the rare DX of Sable Island!

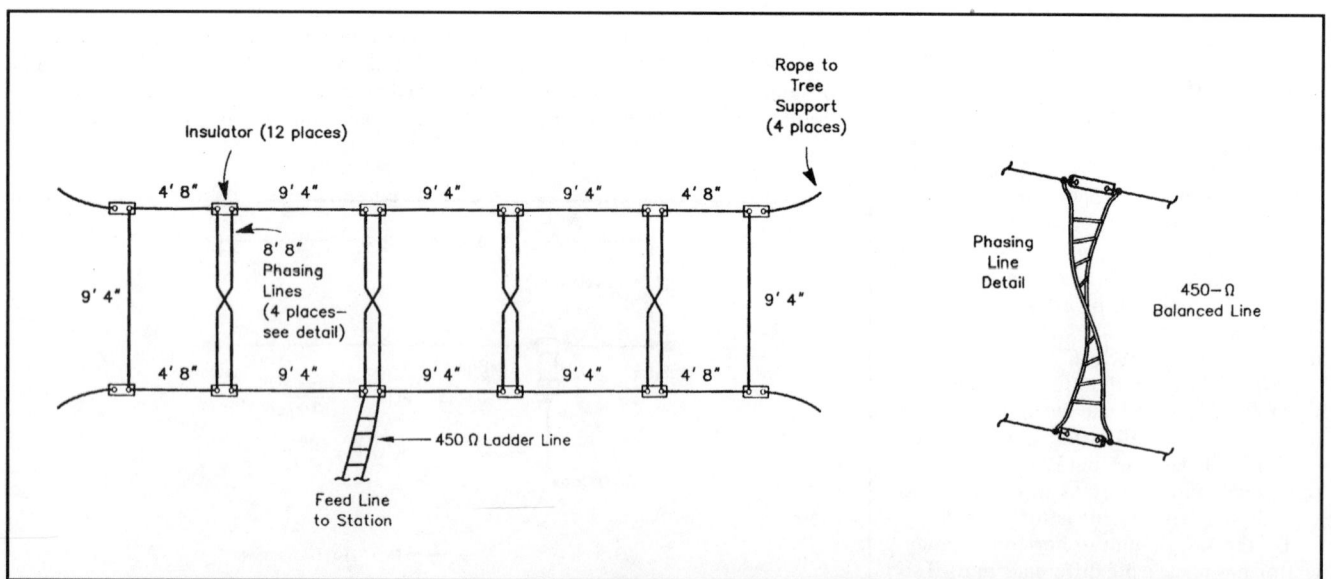

Figure 3—Construction details for an eight-element 6 meter Sterba curtain. The design frequency is 52 MHz. Note that the phasing section is twisted once, so that the conductors cross. The inner end of an upper element feeds the outer end of a lower one. If you have a 6 meter antenna tuner with a balanced output, you can feed the curtain with 450-Ω line between the antenna and the tuner. Otherwise, use a 4:1 balun at the feed point (as in Figures 1 and 2) and you can feed the curtain with 50-Ω coaxial cable. As the name implies, you hang this antenna vertically, just like a window curtain!

to have unequal currents in some of the $1/2$ wavelength sections, but it is still better than the long wire.

The Sterba Curtain

The Sterba curtain antenna is also composed of $1/2$ wavelength radiating sections and phasing sections. An implementation of this antenna for 10 meters was described in the October 1991 *QST* ("Curtains for You," by Jim Cain, K1TN). The antenna shown in Figure 3, made up of eight $1/2$ wavelength elements, should have a gain of about 8 dB over a dipole. This is a physically complex antenna, but it is a superb performer. It is also less than 38 feet long on 6 meters. Again, remember that while enjoying that gain in the most-favored directions, other directions will have less gain.

I haven't tried the Sterba on 6 meters myself, but I'm up for the challenge. Regardless of the complexity, it still beats the cost of a beam and a tower!

Chapter 2
144 MHz

A Compact Two-Element, 2-Meter Beam	Bob Cerreto, WA1FXT	2-1
The Mini-Five Beam	Lee Aurick, W1SE	2-5
Build Your Own 2-Meter Beam	Dale Botkin, N0XAS	2-7
A "Rope Ladder" 2-Meter Quagi	Jim Ford, N6JF	2-9
Try a "Dopplequad" Beam Antenna for 2 Meters	Keith Kunde, K8KK	2-12
A Five-Element 2-Meter Yagi for $20	Ron Hege, K3PF	2-16
A 2-Meter Phased-Array Antenna	Harold Thomas, K6GWN	2-20
7 dB for 7 Bucks	Nathan Loucks, WB0CMT	2-23
Build a Portable Ground Plane Antenna	Zack Lau, W1VT	2-25
Constructing a Simple 5/8-Wavelength Vertical Antenna for 2 Meters	J. Bauer, W9WQ	2-27
A Glass-Mounted 2-Meter Mobile Antenna	Bill English, N6TIW	2-29
Build a Weatherproof PVC J-Pole Antenna	Dennis Blanchard, K1YPP	2-33
The W3KH Quadrifilar Helix Antenna	Gene Ruperto, W3KH	2-36
Try Copper for 2 Meters—The Cu Loop	Dick Stroud, W9SR	2-40
A Five-Element Quad Antenna for 2 Meters	Jim Reynante, KD6GLF	2-43
Recycling TV Antennas for 2-Meter Use	Ronald Lumachi, WB2CQM	2-46
A Portable Quad for 2 Meters	R.J. Decesari, WA9GDZ	2-49

By Lee Lumpkin, KB8WEV, and Bob Cerreto, WA1FXT

From *QST*, January 2000

A Compact Two-Element, 2-Meter Beam

What do you get when you take an already-unusual design for an HF beam antenna, scale it to VHF and turn it on its ear? A vertically polarized modified Moxon, of course! Build this wire and PVC beauty to solve your 2-meter troubles in a jiffy.

I'd been looking for an antenna to monitor 2-meter simplex and Skywarn frequencies that was affordable and easy to install in my attic. Bob, WA1FXT, and I live in an area that sees Skywarn activations for tornado and severe thunderstorm watches several times a year. I also live in a house where my shack is in a new addition, separated from the rest of the house. When I'm in the older part of the house, I have a much better view of weather approaching from the west and north (the usual directions), but I can't hear the radios in my shack. I'm also out of touch with local 2-meter simplex frequencies when I'm not near the radio room. Bob has a similar situation at his home.

Finding the Design

Bob and I had been discussing HF and 6-meter Field Day antennas. One day, I visited L.B. Cebik's (W4RNL) Web site at: **www.cebik.com**. The site is an excellent place to find antenna information and it's a valuable resource for those educating themselves about antennas. While considering his refinement of HF beams designed by Les Moxon, G6XN, I realized that these interesting gain antennas had the characteristics I considered ideal for a 2-meter attic antenna—smooth, wide front lobes with no notches, reasonable gain, relatively compact dimensions and ease of construction and feeding.

Modifying the Design

I cut and pasted W4RNL's dimensions for horizontally polarized HF wire Moxon beams into a spreadsheet and derived formulas for the dimensions. I oriented the antenna to achieve the vertical polarization needed for 2-meter FM and took the 10-meter dimensions and put them into Roy Lewallen's *EZNEC* antenna design and analysis program. Using the formulas I had derived from Cebik's plans, I rescaled the antenna for 2 meters and tweaked it a bit to overcome the large shift in the element length-to-diameter ratio. The resulting design characteristics contained a pleasant surprise. The single, smooth front lobe widened to about 135° along the horizon (the –3dB beamwidth). This vertically polarized variant was much broader than its horizontally polarized cousin.

While tracking how the antenna's pattern changed at several points in the 2-meter band, I found a point about 500 kHz above the design frequency that had a single rear notch in the gain pattern at the cost of marginally higher SWR. This resulted in a cardioid-type pattern with a relatively narrow notch to the rear that was about 35 dB down from the maximum forward gain (which models at around 6 dBi in free space). Far from being a disappointment, this was a useful foxhunting antenna

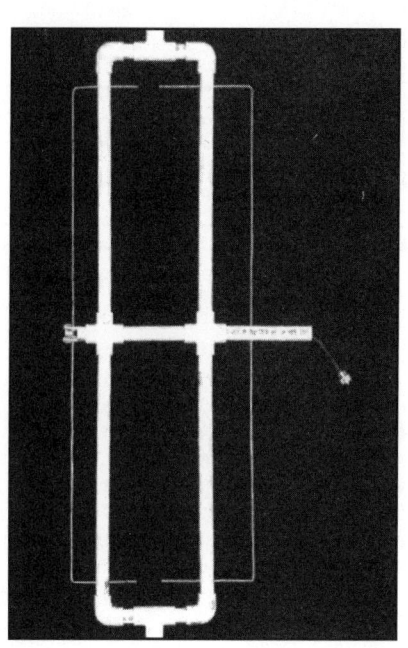

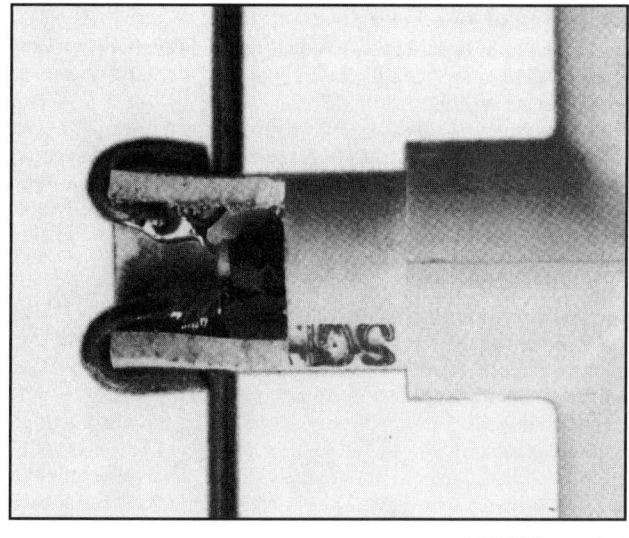

A closeup view of the connection between the coax and the radiating element.

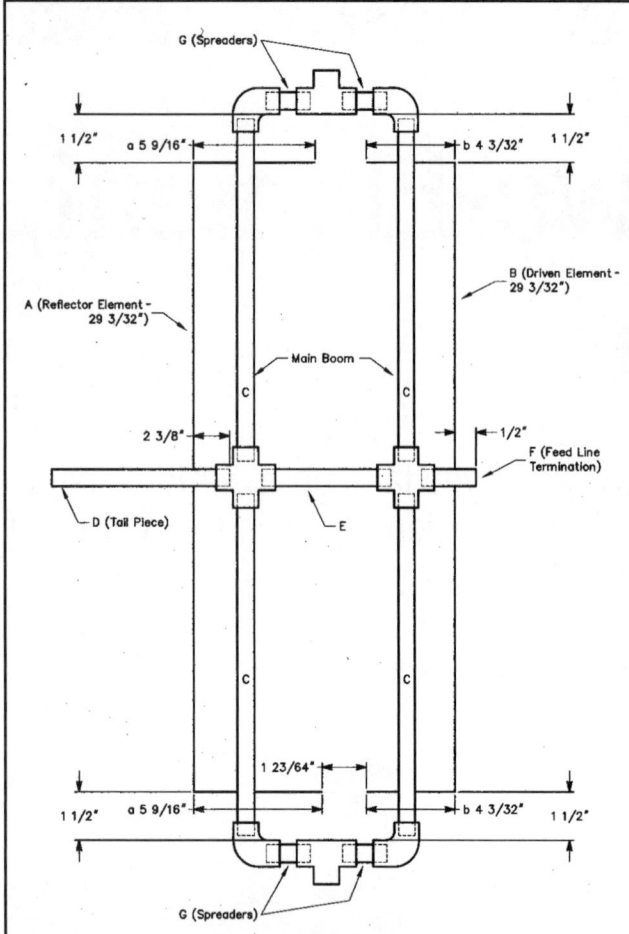

Figure 1—A construction diagram of the Moxon 2-meter beam antenna. See text and tables for details.

Table 1
Bill of Materials

10 feet	½-inch Schedule 40 PVC pipe
4	½-inch Schedule 40 PVC 90° elbows
2	½-inch Schedule 40 PVC crosses
2	½-inch Schedule 40 PVC tees
Approx 10 feet	No. 10 AWG copper wire
1	PL-259 coax connector
1	UG-176U coax adapter
2	Wire ties
3	Amidon Ferrite Beads (FB-43-5621) Amidon Associates 240-250 Briggs Ave Costa Mesa, CA 92626
RG-8X coax	As needed
Misc	Plastic electrical tape

PVC and Wire Cutting Schedule

Reflector element **A** consists of one piece of #10 AWG copper with a straight section and two tails bent at 90°. Total length is a single piece of $40^{7}/_{32}$ inches.

Driven element **B** is two half elements fed in the middle. Total length is a nominal $37^{9}/_{32}$ inches, but build it according to the text. A slightly longer wire is required to wrap around the PVC and secure the feed point to the pipe. See Figure 1 for the drawing labels.

Wire

Qty	Label	Description	Length
1	A	Reflector Element	$29^{3}/_{32}$"
	a	Reflector Element Tail	$5^{9}/_{16}$"
1	B	Driven Element	$29^{3}/_{32}$"
	b	Driven Element Tail	$4^{3}/_{32}$"

PVC

Qty	Label	Description	Length
4	C	Main Boom	$15^{9}/_{16}$"
1	D	Tail Piece	6"
1	E	Middle Boom Spreader	$5^{1}/_{8}$"
1	F	Feedline Termination	2"
4	G	End Boom Spreader	$2^{1}/_{16}$"

of manageable size. This was something Bob and I had both been looking for!

We needed a framework to support the wire elements, so I put the dimensions for the two wire elements into a CAD program and worked up a PVC framework to support them. When Bob saw the antenna pattern and construction plans he became excited about the antenna's possibilities—especially as a foxhunter.

Build One of your Own

Please note that the dimensional accuracy used here is overkill. If you cut your elements to within an eighth of an inch, your antenna should work like a champ, with no practical loss in performance. Put away the calipers and pick up a ruler! The most critical dimension of these "modified Moxons" is the distance between the tips of the reflector tails and driven elements, which can be fine-tuned after assembly.

The first step in construction is to cut the ½-inch schedule 40 PVC pipes to the lengths needed for the support frame. The quantities and lengths are listed in the sidebar. If you're going to vary from the pipe lengths shown in the table, be sure to allow ¾-inch to accommodate the various PVC fittings.

After cutting the PVC to length (per the accompanying table), measure and drill the holes for the wire antenna elements. Use a drill press and a fence to center the holes in the pipe (or mark the holes and drill carefully by hand).

Place the tip of a center punch or a nail at the drilling point and tap with a hammer to dimple the surface. This will hold the tip of the drill bit in place and keep it from spinning off target as you drill.

Follow the diagram in Figure 1 as you begin to assemble the antenna. Drill a hole completely through each of the four pipes labeled **C**, 1½ inches from one end. This will go at the end farthest from the antenna center. The hole drilled in pipe **F** should pass all the way through, ½-inch from the pipe end. Mark one end of pipe **D**; that end that will be inserted into the PVC cross. Drill a hole completely through pipe **D** 2⅜ inches from the marked end. Make sure to orient the pipe properly when assembling the PVC frame or you will end up with a bowed reflector element (**A**).

Use solid #10 AWG copper wire for the antenna elements. Number 10 copper wire is a nominal 0.1-inch in diameter, but ours was a bit too large for a $^{7}/_{64}$-inch hole, so we used a ⅛-inch drill bit. This allows the wire to pass through easily without too much slop. If your wire slips through the holes after building the antenna, hold the elements in place with wire ties, heat shrink tubing, hot glue, RTV sealant or tape (anything that won't detune the antenna). If you're making a permanent outdoor antenna, use UV-resistant material to secure the elements.

The next step is to assemble the PVC frame. This is best done on a flat surface. Use a rubber or wooden mallet to persuade the pipes to seat snugly in the fittings, being careful to keep the wire holes in the correct plane.

Align the four vertical pipes (**C**) by sighting through the holes you drilled in each pipe for wire tails **a** and **b**, aligning them with the corresponding holes in the opposite vertical pipe. If necessary, a pair of pliers can be used to fine tune the alignment by twisting the pipe in the fittings. The front and tail pipes, **F** and **D**, are aligned by sighting along the lengths of the vertical support pipes to be sure that the holes for wire sections **A** and **B** run parallel with the frame.

Next, prepare and install the wire elements. Straighten some #10 wire before starting the installation. After inserting the

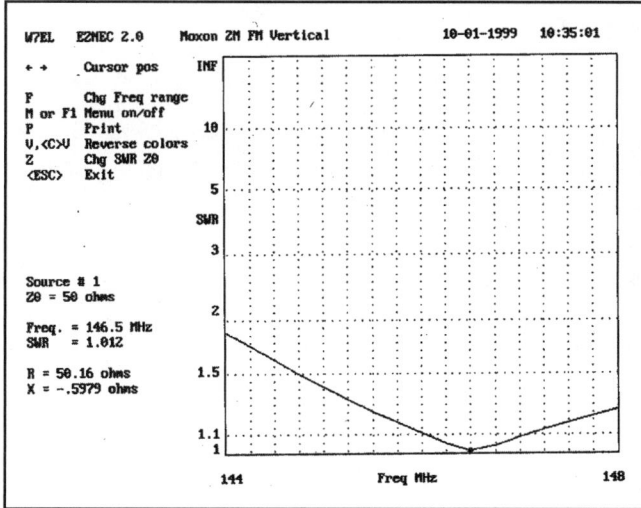

Figure 2—The predicted SWR bandwidth of our 2-meter beam, from 144 to 148 MHz.

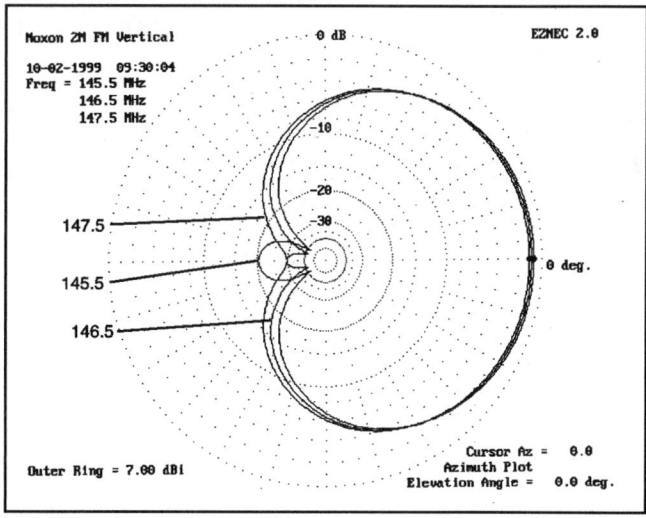

Figure 3—The predicted radiation pattern of the Moxon beam.

wire into the PVC frame, it's difficult to thoroughly clean the wire at the feed point (**F**), so be sure to clean it thoroughly beforehand. Oxidized wire makes it difficult to get a good solder joint, especially in this tight space. So use an abrasive plastic pad or fine steel wool to clean up the wire ends at the feed point to make for easier soldering.

First, install one half of the driven element (**B**). Take a longer piece of #10 wire (about 24 inches) and bend an inch at one end about 90°. Do this at the end you have already cleaned in preparation for soldering at the feed point. Bend about half of that 1-inch bend another 30° to 45° in the same plane. Pass the resulting "J" shape through the feed point hole in PVC pipe **F** from the outside (so the bends end up inside the pipe with about 1/2-inch protruding from the open end of the pipe). Hold the wire against the inside of the PVC pipe with pliers or a dowel and wrap the 1/2-inch of protruding wire back around the outside of the pipe and crimp it tightly against the outside of the pipe. See the close-up photo of the feed point to see the final result.

Once this is done, hold the wire parallel to the front vertical portion of PVC support frame **C** and bend it back at 90° at a point in line with the support holes in PVC pipe **C**, forming element tail **b**. This will be about 14 1/8 inches from the outside of PVC pipe **F**.

Measure and cut the bent-back wire tail section (**b**) to 4 3/32 inches long. Leave extra and trim later if you wish. Pass this tail section through the support hole in the front vertical pipe (**C**) on the PVC frame. Install the other half of driven element **B** using the same steps and dimensions.

Now you need to feed the antenna with coax *before* installing the reflector element (wire **A**). Prepare the feed line by obtaining a three-foot RG-8X jumper (with PL-259s installed on each end) and cutting it in half. This speeds construction and makes it easy to build two antennas at once. You can, of course, make up your own RG-8X coaxial cable and PL-259 assembly.

Strip 1 inch of insulation from the shield and 3/4-inch from the center conductor at the unterminated end of the coax. Use three type-43 ferrite choke beads (Amidon FB-43-5621) to keep RF from returning along the coax shield. Slip the beads onto the feed line and secure them with wire ties and electrical tape as close to the stripped end of the coax as you can without touching the bare copper braid, the center conductor or the wire elements at the feed point. The electrical tape should also help prevent this contact.

To improve access to the feed point, cut a section out of the side of the pipe at the feed point as shown in the photo. First cut into one side of pipe **F** just behind the feed point, perpendicular to the length of the pipe to a depth of about 1/3 of the pipe diameter. Another cut in from the end of pipe **F**, along its length, but cutting in only about 1/3 its diameter, leaves enough support for the center of driven element **B**. This leaves plenty of space for soldering.

We considered making these cuts before installing the driven element wires, but decided that the stresses involved in bending the heavy gauge wire around the PVC at the feed point might cause cracking. You may find that it's okay to make your access cut first. Be sure to use a fine blade, such as a hacksaw or dovetail saw.

Now run the feed line in from the rear of the center horizontal support pipe up to the feed point, through pipes **D**, **E** and **F**. Wrap the coax shield around one side of driven element wire **B** (inside pipe **F**) and the center conductor around the other end of the driven element wire, then solder both connections. Use paste flux and a 100-W iron of sufficient mass. Irons with less thermal capacity can't generate enough heat, or have their thermal energy conducted away too quickly by the #10 wire. Lengthy heating with a smaller iron is likely to melt the PVC.

Now it's time to put reflector element **A** into place. First, pass 45 inches (or more) of #10 wire through the rear center support holes in PVC pipe **D**, being careful to go around the coaxial feed line rather than through it. Bend about six inches of wire at one end back at 90° toward the driven element and support frame (forming tail **a**) and trim it 5 9/16 inches from the bend. Then, pass this tail through the wire support holes in the rear vertical section (**C**) of the PVC support frame. Measure 29 3/32 inches from this first bend along reflector wire **A** and make the bend to form the other reflector tail. Trim this tail to 5 9/16 inches and pass it through its support hole in pipe **C**.

Everything is now in place, so square up the wire elements on the frame. One characteristic of Moxon antennas is a sensitivity to the relative positions of tails **A** and **B**, so make sure the tails are in line with each other and spaced at 1 23/64 inches. This was considered in designing the PVC support frame and the points at which it holds the wire elements. This design allows the wire tails to be held in line with each other, leaving the distance between the tips of the tails to be fine tuned and then taped, glued or otherwise secured to the PVC frame once the antenna is performing to spec.

We didn't cement our PVC frames because my antenna would be installed indoors and the joints were firmly seated without gluing. If you want to cement yours, we'd suggest assembling the frame to align the wire support holes and carefully making reference marks at the junctions of the pipes and fittings. This will allow you to quickly orient the PVC elements before the PVC glue sets up. If you're going to mount your antenna outside, gluing the PVC frame is a good idea. PVC glue sets up very quickly, so if you don't feel confident, you might want to insert small sheet metal screws into predrilled holes instead. Alternatively, you can build the PVC frame and glue it together before drilling the wire support holes.

You should try to run the feed line away from the antenna for a couple of feet before running it parallel with the main sections of the wire elements. Running it parallel to these sections at less than 19 inches or so may distort the pattern of the antenna and change its SWR.

Performance Testing

After building the first antenna, we decided to test its performance before building the second. We took the antenna to a clear spot in the yard and hooked it up to an MFJ-259 antenna analyzer through about 25 feet of RG-213 coax. Bob held the antenna up on a PVC mast while I ran through 2 meters with the analyzer—and I started to laugh.

Bob was dying with curiosity, so we traded places while he swept the band. The antenna came up on the frequency we expected, with an even broader bandwidth (see Figure 2). We decided to check the front-to-back ratio, but had no field strength meter at hand. Bob got in his truck and I aimed the front lobe of the antenna at him until his receiver dropped below S-meter saturation about a mile away.

While I turned the antenna and reported to Bob where he was in the pattern, a station in my normal fringe area called me. This station doesn't receive me full quieting on my base antenna (stacked 5/8-wave omni antennas at 33 feet, fed with about 2 W), but was hearing me now with the Moxon up only eight feet while running 350 mW!

Bob and this other station (11 miles in the opposite direction) reported S-meter readings in line with the computer-predicted pattern as I rotated the antenna in azimuth. At this point, we scrambled back to the garage to build a second antenna before we ran out of time!

The Hunt

About a week later, Bob and I got together for a foxhunt and to check out the front-to-back ratio in a more controlled manner. Bob, his son Matt, and I were on our first hunt together. Bob had borrowed a passive field-strength meter, but we were unsure of its linearity and were unable to get enough useful range out of it to measure the front-to-back ratio on the 2-meter Moxon.

Bob has a well-calibrated attenuator, so we made some measurements and checked the pattern and front-to-back ratio by switching the attenuator to give the same reading on the field-strength meter. The pattern turned out to be in line with the computer model, and the narrow rear notch was down 29 dB from the front lobe, exactly as predicted. We also checked the antenna at 147 MHz, about 500 kHz above the design frequency of 146.5 MHz. Again, as predicted, the rear notch deepened to about –35 dB.

We hunted using the rear notch in the Moxon pattern. We were thrown off several times by reflections from large metal buildings, a power plant that killed the signal, and the unfamiliar terrain. But we overcame the problems and were the third team to find the fox. We covered 19 road miles in the search for the fox, which was about six miles from the start as the crow flies. On one transmission, the fox was on a vertical omni with steady power when we were a couple of miles out. This gave us our best sample. The bearing we took on this transmission was within a couple of hundred feet from the fox's actual position.

The hunter who first found the fox (in half our time) and the team that found it a few seconds ahead of us were both using the same 4-element, balanced feed Yagi that I had used to win two California foxhunts. When tuned properly this Yagi has a single rear notch. The Moxon performed similarly, with the exception of its wider front pattern and reduced forward gain. We were very happy with its performance as a compact foxhunting antenna.

After we had designed and tested this antenna, we decided to see how it would act mounted on a standoff from a metal mast. We haven't checked thoroughly, but computer modeling suggests that at $1/4$ wavelength, the rear null is not nearly as deep. The cardioid notch pretty much disappears. At $1/2$ and 1 wavelength, the pattern is pretty good, so choose your mounting offsets (from a vertical metal mast) accordingly. This was not a great concern to me because I used PVC tee connectors at the top and bottom of the frame to support the antenna with short sections of PVC pipe attached to the ridge beam and a ceiling joist in my attic. It has performed very respectably there, only 12 feet off the ground. I can make all of my regular 2-meter simplex contacts, and the antenna holds its own when accessing local repeaters.

Resulting Antenna Performance and Potential Uses

No antenna does everything well, but this design has a number of useful characteristics:

• A smooth, wide front lobe (see Figure 3) with modest, but useful gain (of about 6 dBi in free space) and none of the sidelobe notches associated with most Yagi and quad beams of three or more elements.

• A single, deep rear notch (up to –35 dB relative to the maximum front lobe gain). This makes it useful for rejecting single-source interference and for foxhunting.

• A compact and simple design that is inexpensive and easy to build with minimal tools and skills.

• A very good direct match to 50-Ω feed lines.

You can use this antenna to: minimize or eliminate interference or intermod from pagers or other stations while still receiving desired signals from most other directions; access desired repeaters while rejecting an unwanted repeater; and reach a broad swath of stations or repeaters with reasonable gain and no need to rotate a beam or overcome multiple side lobe nulls that accompany multi-element Yagis and quads. You should also be able to foxhunt by placing the fox's signal in the single null and heading in the direction of greatest signal attenuation.

It should be noted that L. B. Cebik is responsible for refining the geometry of the Moxon beam to its full potential. Our antenna is a simple rescaling of his work. He is very generous in sharing his work with anyone who is interested. His work on this antenna was inspired by the designs of Les Moxon, G6XN, and Fred Caton, VK2ABQ, who started out with square HF wire beams using buttons to insulate the element tails.

This particular version of the Moxon antenna should be used as a starting point. Visit Cebik's web site for a horizontal 10-meter version made from aluminum tubing. Cebik has also suggested using the Moxon's wide front lobe to point directly toward the zenith for "unsteered" satellite communication on 2 meters—a use which deserves attention and development.

Bibliography & Resources:

www.cebik.com. Learn from L. B.'s experience and modeling expertise. This site is a wonderful resource for antenna experimenters.

HF Antennas for All Locations, 2nd ed., Les Moxon, RSGB, ISBN 1-872309-15-1

"Moxon Rectangles for 40-10 Meters," L. B. Cebik, *QRPp,* Dec 1995, pp25-27.

"An Aluminum Moxon Rectangle for 10 Meters," L. B. Cebik, *Antenna Compendium*, Vol. 6 (ARRL).

"The Moxon Rectangle on 2 Meters," *AntenneX,* Sep 1999, L. B. Cebik.

"Building a 2-Meter Moxon," *AntenneX* (Oct 1999), L. B. Cebik.

"Moxon Rectangles: A Review," *AntenneX,* Oct 1998, L. B. Cebik.

"Modeling and Understanding Small Beams: Part 2: VK2ABQ Squares and the Modified Moxon Rectangle," *Communications Quarterly*, Spring 1995, pp 55-70.

"The Moxon Rectangle," Morrison Hoyle, VK3BCY, *Radio and Communications* (Australia), Jul 1999, pp 52-53.

By Lee Aurick, W1SE

From *QST*, May 1997

The Mini-Five Beam

A compact 2-meter beam antenna that's ideal for attic installations!

Most 2-meter beam antennas use half-wavelength elements. If you want to install one of these antennas with vertical polarization for FM operation, you can do it in one of two ways. You can secure the end of the boom—at a point just behind the reflector—to a metal mast. Or, you can attach a nonconductive (wood or PVC) mast to the middle of the boom. If you're installing the antenna in the great outdoors, you have plenty of room to accommodate the lengths of the elements. In most cramped attics, however, you won't stand a chance. Wouldn't it be nice if you could somehow cut the elements in half? Hmmm....

The Solution

What follows started out as an experiment to answer the question, "Why not?" and it succeeded so well that I wanted to share this novel approach with other 2-meter FM enthusiasts. The solution was so simple, I don't know why I didn't see it before. That is, why use half-wavelength elements? Why not use $1/4$-wavelength elements mounted entirely above a metallic boom? The height (for vertical polarization) of such an antenna would be less than 2 feet.

Just as a quarter-wavelength vertical antenna will "see" its other half in the groundplane on which it is mounted, I reasoned that quarter-wavelength elements would see their opposite half in the metal antenna boom itself. The big question was: Would the gain and directivity resemble a beam with half-wavelength elements? Read on....

Element Construction and Mounting

The driven element and its gamma-match rod were cut from $1/4$-inch-diameter copper tubing. I could have used aluminum, but copper is much easier to solder. Hard-drawn aluminum wire, available from Radio Shack, is used for the other four elements. The boom is made of a 72-inch length of 1-inch-square aluminum stock.

The only tedious part of building this antenna was straightening the aluminum wire (it comes on a roll). This was accomplished by placing an oversized piece of wire (perhaps three inches more than was needed) between two boards, clamped in a vise. Working the wire back and forth between the boards, while tightening the vise, results in a reasonably straight piece of wire.

Using a pair of needle-nose pliers, make a small loop at one end of each of the aluminum wire elements. Then make a 90° bend in the wire immediately adjacent to the loop you just formed. This loop is used to pass a self-tapping screw to mount the elements to the top of the boom. After making the small loop in the wire, cut the reflector element to 20 inches and the three directors to 18.5 inches each, as measured from the free end of the wire to the start of the 90° bend.

Now let's go to the driven element. Use a hammer to flatten a 1-inch length of the tubing at one end, and make a 90° bend, so the flattened part can be used as a mounting bracket. Drill two holes so that you can attach it to the boom with self-tapping screws. Measure 19 inches from the 90° bend and cut off the excess tubing.

Attach the elements to the boom using the spacing dimensions shown in Figure 1. Use toothed lockwashers under all screws, and between the elements and the boom.

Building and Installing the Gamma Match

To match the beam to 50-Ω coaxial cable and achieve a low SWR, you need to find a way to electrically "tap" the proper point on the driven element. That's the function of the gamma match. While the gamma match allows you to "move" along the element to get the required match, its presence introduces inductive reactance. This will prevent you from achieving the lowest possible SWR. You cancel this inductive reactance by introducing an equal, but opposite, capacitive reactance. All you need to work this magic is a tiny capacitor.

Find a small plastic container, such as a 35-mm film canister, in which to mount your gamma capacitor (see Figure 2), a 45-pF variable "trimmer" capacitor (Mouser Electronics part no. 242-3610-45; tel 800-346-6873). Drill two holes in the

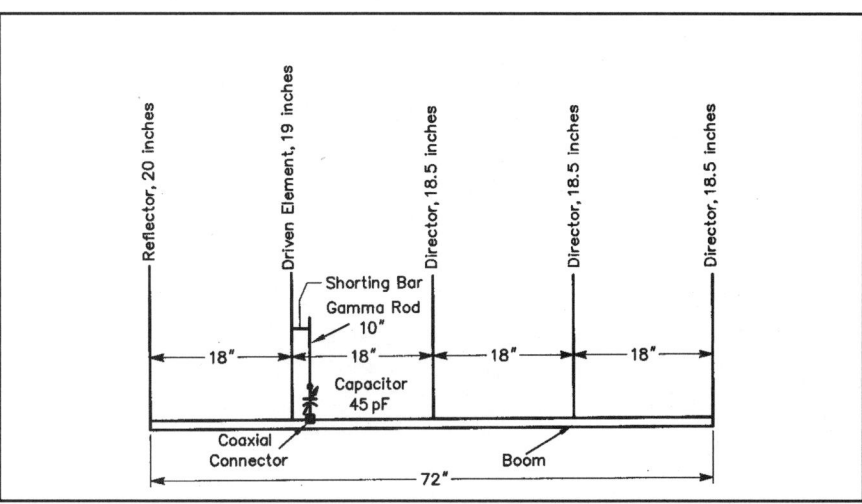

Figure 1—Assembly drawing of the Mini-Five beam antenna.

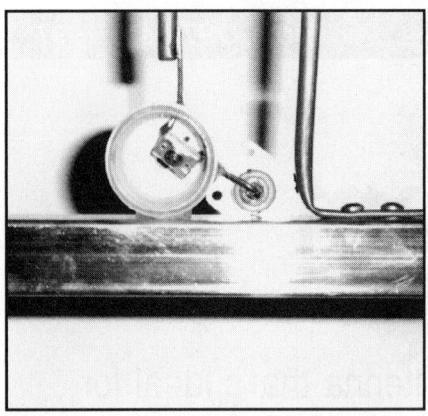

Figure 2—Close-up view of the gamma capacitor within its plastic housing. One wire attaches to the gamma rod while the other is soldered to the center pin of the coaxial connector. Note that the coaxial connector is attached to the boom using a single self-tapping screw.

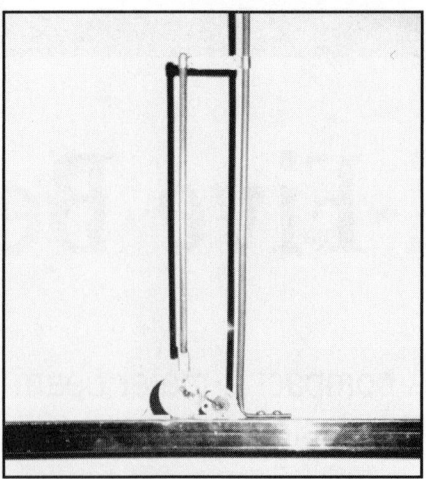

Figure 3—The complete Mini-Five gamma match assembly. Tune the antenna for the lowest SWR by sliding the shorting bar along the rod (for coarse tuning) and adjusting the variable capacitor (for fine tuning). See text.

sides of the plastic container, just large enough to pass a 2-inch length of stiff wire through each hole. Solder the capacitor to the ends of these wires, and place the capacitor inside the container. Apply a drop of silicon sealant to the wires where they penetrate the container.

Solder one wire to the center conductor of the coax connector, after you secure the coax connector to the boom by putting a self-tapping screw through one of its mounting holes. Solder the other wire to the end of the gamma rod, and use silicon sealant to glue the container to the boom.

The gamma rod is merely a 10-inch length of copper tubing mounted about 2 inches away from the driven element. To hold the bar in place (and to connect it electrically to the driven element), you need to make an adjustable shorting bar (see Figure 3). Make the bar from a $1/32$-inch-thick, $3/8$-inch-wide aluminum strip about 5 inches in length. Bend each end around a piece of scrap from the $1/4$-inch-diameter copper tubing used for the driven element and gamma rod. Drill a hole in each overlapping portion of the strip and install #4 machine screws and nuts through each hole to clamp it to the gamma rod and driven element. Slide the shorting bar over the driven element and gamma rod, and finger-tighten the screws.

Fasten a piece of aluminum stock, 5 inches square by $1/16$-inch thick to the boom with large self-tapping screws. Drill holes through this piece for the **U** bolts (also purchased at Radio Shack). The **U** bolts may be used to mount the boom to any available mast.

Tune Up

If there is anything I like to do, it's tune up an antenna. There is an undeniable joy in seeing metal and wires spring to life. It was a mixed blessing, then, when the beam was perfectly matched in less than 10 minutes. It meant that the tuning and tinkering, so dear to the heart of every antenna experimenter, was over!

Initially, I adjusted the shorting bar to be about 7 inches above the boom. The SWR meter indicated a minimum SWR of about 1.5:1 as the gamma-capacitor was adjusted through its range—not bad for a first cut.

I should point out that your body has a marked effect on SWR readings; a reminder that antennas of all types like to be by themselves, and detest the presence of humans, trees, and other conductive objects. Try to keep as much of your body as possible below the level of the boom, and tune the capacitor at arm's length, using a nonconductive tool. It helps to have a friend at the radio during this procedure.

I loosened the screws and moved the shorting bar to a point about 8 inches above the boom, and swept the capacitor through its range. The meter showed a minimum SWR of about 1.3:1—a definite improvement. Then I moved the shorting bar about 9 inches above the boom and, with a joyous heart, read an SWR of about 1.1:1.

But the purist in me cried out for one more effort. This time I moved the shorting bar just $1/2$ inch, for a total of $9 1/2$ inches from the boom. The capacitor was again exercised, and there it was: a 1:1 SWR.

The Acid Test

How would this antenna function in the real world? No matter how easy it was to construct, or how well it tuned, if it didn't cut the mustard by providing the muscle to reach distant repeaters, it wouldn't be worth much as an antenna.

I have a weekly sked with an old friend who is about 50 miles away. We use a repeater that is about 35 miles from my location. My primary antenna, two collinear $5/8$-wavelength elements at 30 feet, does an adequate job, but noise is occasionally a problem. Would my new Mini-Five beam make an improvement?

The beam was mounted on its 10-foot test mast in the drive-way and pointed toward the repeater. I squeezed the mike button and signed with my call sign, almost apologetically. When I released the button I heard a loud CW identification. The repeater was full quieting in my receiver!

Okay, so the Mini-Five has some gain. How about directivity? Using other repeaters as signal sources, I rotated the antenna and noted the signal strengths. The beam appeared to be directive, indeed, with about a 45° beamwidth between the half-power points.

Conclusions

The experiment with this unique beam antenna certainly appears to have been successful. The Mini-Five's performance is very satisfactory. It is smaller and lighter, too! The Mini-Five has much to offer hams who use low-power rigs and live in condos, or in homes in which outside antennas are not permitted. The 20-inch height and three-foot turning radius makes it possible to install this antenna, *and a rotator*, in the smallest attics!

By Dale Botkin, NØXAS

From *QST*, October 1995

Build Your Own 2-Meter Beam!

All you need is a few dollars and a couple of hours to build this simple directional antenna.

If you're like most new hams, you probably own a 2-meter hand-held FM transceiver (H-T) that came equipped with a flexible "rubber duck" antenna. These antennas are fine for many portable and mobile applications, but they often perform like rubber dummy loads when you need greater range. If you live far from your favorite repeater, like I do, you probably have to use your H-T's highest power setting to reach it—assuming you can reach it at all.

The solution is a better antenna, but what kind? I wanted to chat with a friend who lives in another town. Not only was my rubber duck completely inadequate, my homebrew J-pole antenna didn't help, either. What I needed was an antenna that would focus my H-T's meager output in a particular direction. In other words, I needed a *beam*.

Too Difficult?

Like most people, I had always assumed that beam antennas were tricky to build and tough to tune. I also figured that any beam with less than three or four elements wasn't worth my time. Most of the ads and plans I saw were for big, 9 to 22-element antennas with tons of gain.

You can certainly spend a lot of time designing, building, and optimizing a large beam antenna from scratch. (You'll at least save some money!) But I discovered that a smaller, simpler beam offers more than enough performance for casual operating. The big antennas are great if you're trying to work 2-meter SSB, satellites and so on. To expand the range of your FM voice or packet station, however, this easy *quad* antenna is just the ticket!

After finding plans in the *ARRL Antenna Book* for a two-element quad antenna, I made a quick trip around the house to see what materials I had on hand. Because we had recently completed a new deck, I had lots of scrap wood, including some pieces that were about $1^{1}/_{2} \times {}^{3}/_{8}$ inches and a little under two feet long. They were perfect for the job. I did have to make a trip to the hardware store for some #12 copper wire (I removed the insulation from #12 electrical wiring). You could use #8 aluminum wire instead (clothesline wire), but it is much more difficult to solder. Using larger wire, or insulated wire, will affect the resonant frequency of the antenna. The dimensions described in the text are for #12 bare wire.

Construction

Cut one piece of wire to $85^{1}/_{2}$ inches in length. This will become the *reflector* element. Make a 90° bend in this wire $10^{5}/_{8}$ inches from one of the ends, then measure $21^{3}/_{8}$ inches from this bend and make another 90° bend. Measure another $21^{3}/_{8}$ inches and make another bend, another $21^{3}/_{8}$ inches, another bend. You'll eventually wind up with a square $21^{3}/_{8}$ inches on each side (see Figure 1). Solder the ends of the wire together to complete the square.

Now cut another piece of wire to $81^{1}/_{2}$

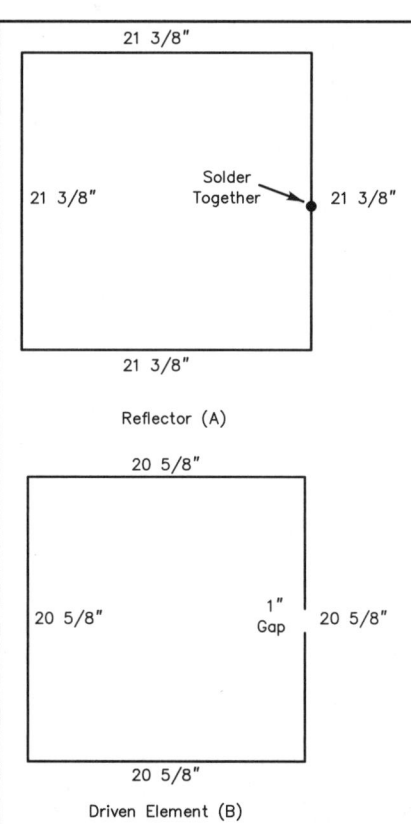

Figure 1—When you're finished bending the reflector wire, you should have a square $21^{3}/_{8}$ inches on each side (A). The driven element is $20^{5}/_{8}$ inches on each side with a small gap (B).

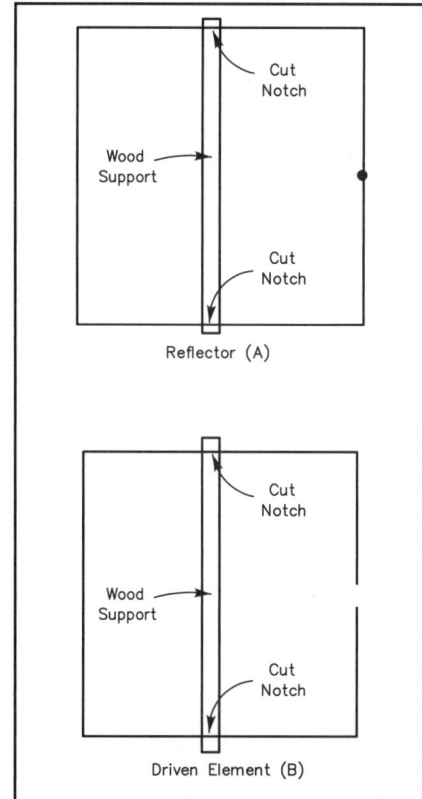

Figure 2—Lay the wood supports on top of the wire loops and draw lines where the wires cross the supports. Cut deep notches along the lines and use epoxy cement to glue the wires into the notches.

144 MHz   2-7

inches in length. Bend it as I've just described, but use 20⅝ inches for each side. Start 9⅞ inches from the end to make sure you wind up with the ends in the middle of one side, just like the other loop. You'll wind up with a gap of an inch or so between the two ends (see Figure 1). *Don't* solder the ends together! This is the driven element and the gap is the point where we'll attach the coaxial cable.

Now for the supports. Lay one wood support across each loop (see Figure 2) and mark the points where the wires cross. Cut notches in the supports and slide the wires into the notches. Center the supports in both loops. For the driven element, make sure the wood support is *parallel* to the side where the coax will attach. Use epoxy cement to hold the wires in the notches.

When the glue dries, the loops can be attached to the boom (the third piece of wood stock). Secure the driven element (the open loop) at one end with screws or glue (Figure 3). Tack the other loop onto the boom about 9 inches behind it. Use small tacks, tape or some other means to hold this loop in place temporarily. You may need to move it soon. (More about this in a moment.)

All that's left is the feed line. Strip back the outer insulation from the last couple of inches of coax and separate the center conductor and shield braid. Solder the shield to one side of the driven element and the center conductor to the other side (Figure 4). Install a BNC connector on the other end of the coax, and connect the coax to your H-T.

SWR Adjustment

If you can get your hands on a VHF/UHF SWR meter that's capable of making measurements at low power levels, take a few moments to adjust your quad. Set up the antenna away from nearby metal objects. Transmit at 146.52 MHz and check the SWR. If the SWR is 1.5:1 or less, secure the reflector loop permanently and leave the antenna alone. If not, move the reflector loop in one direction or the other, stopping to make more measurements. As soon as you reach the 1.5:1 SWR point (or lower), stop. You're ready to fly!

Many factors affect the resonant frequency of an antenna, especially at VHF. The dimensions I recommend were developed using antenna-modeling software, verified by testing in the ARRL Lab. However, your mileage may vary. The type of wood, the way the antenna is attached to the wood and the size of the wire can affect the tuning. The dimensions are fairly critical; a ¼-inch error will change the resonant frequency by about 500 kHz. Most H-Ts are very tolerant of high SWR, so the antenna will work even if it is not perfect. If you do use an SWR meter to measure the antenna's performance, you may want to adjust the length of the driven element slightly to achieve resonance in the part of the band you use most. Lengthen the antenna by ¼ inch to lower the resonant frequency; shorten the antenna to raise the resonant point.

Bear in mind that the signal direction will be toward the driven element. In other words, you want the loop with the coaxial connection to point toward your target. The other loop is called the reflector, because it does just that—it reflects RF energy toward the driven element.

Results

The first time I tested my new antenna, I simply held it over my head and aimed it roughly west from my office. My target was Fremont, Nebraska, where my friend Joe lives. It's about 25 miles distant, over hilly terrain, and my office is in a depression surrounded by paging-system towers, a police station, a fire station, and business-band radio antennas! I've never been able to raise the Fremont repeater with my H-T, even with a J-pole (and it's a struggle even with my 15-W mobile rig).

Dialing in the frequency of the Fremont repeater, I pushed the PTT switch and announced my call. I heard the repeater squelch tail loud and clear! Another ham answered! But wait—I had the frequency wrong! This wasn't the Fremont repeater on 146.76 MHz. I had hit the Lincoln machine, about 20 miles farther out, on 146.67 MHz! I changed frequencies and tried the Fremont repeater. Even at my H-T's lowest power setting, I hit the repeater loud and clear.

If you install the antenna outdoors, treat the supports with a wood preservative or several coats of spar varnish. You'll want to use a secure means to attach the supports to the boom if you expect the antenna to survive high wind gusts. Use low-loss coax if you install the antenna more than 20 feet from your station.

You can point the quad in the direction of your favorite repeater and leave it. If you want to reach stations at other points of the compass, you'll need an antenna rotator. An inexpensive TV antenna rotator is more than adequate for the job. You'll find used rotators selling at hamfest fleamarkets and elsewhere for peanuts.

Conclusion

I've been really impressed with the performance of this little antenna, all the more so since it was so incredibly easy and cheap to build. If you're tired of being the weak signal on the band, or if you want to try to see just how many repeaters you can hit, it's a joy. You can also use it for simplex contacts over a much greater range than you'd ever think possible. It's just amazing what a couple of watts of VHF FM can do with the right antenna!

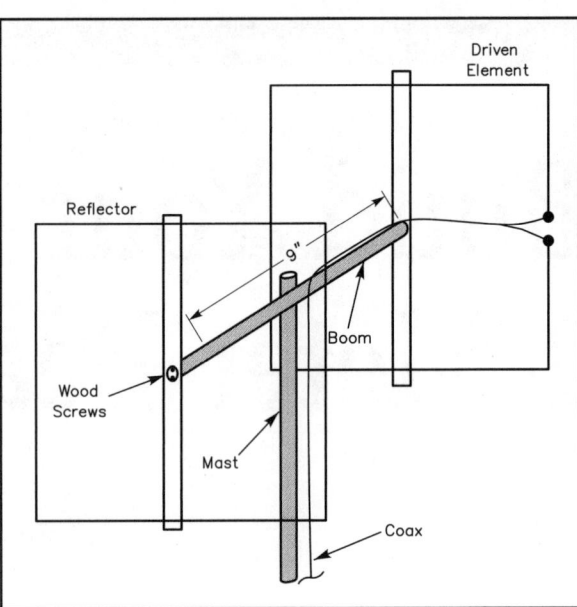

Figure 3—The finished quad should look like this. The boom is attached to either support with two small wood screws. (Other methods can be used as well.) The coaxial cable is routed along the boom and down the mast. The mast can be metal, wood, PVC or whatever suits your fancy.

Parts List

3 1½ × ⅜ × 24-inch wood supports
#12 copper wire (no insulation)
Epoxy cement
Coaxial cable
Small wood screws
BNC connector (Use a PL259 connector [or a BNC-to-PL259 adapter] if you intend to use the antenna with a mobile or base radio.)

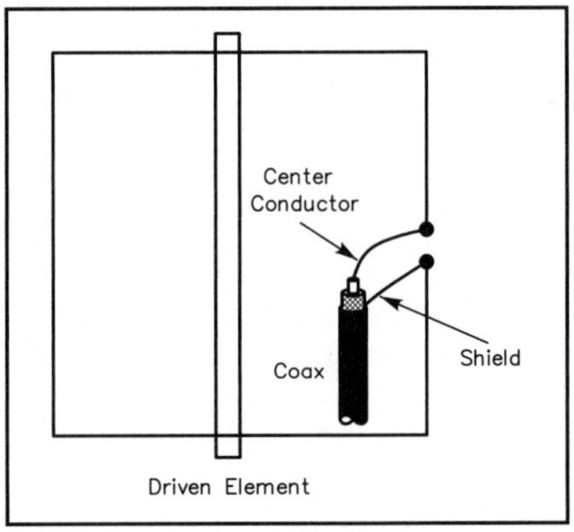

Figure 4—Strip the end of the coaxial cable to expose the center conductor and the shield braid. Solder the center conductor to one side of the driven-element loop and the braid to the other.

A "Rope Ladder" 2-Meter Quagi

A long beam antenna need not be absolutely straight to get out like gangbusters, and this hundred-footer proves it!

In the late 1970s, Wayne Overbeck, N6NB (then K6YNB), came up with a very popular 2-meter antenna called the quagi—a combination of a quad and a Yagi-Uda array.[1,2] He later extended his design to cover other frequencies. Over 20 years ago, A. H. C. Smirk, ZL4TAH, wrote about a long rope-supported 2-meter Yagi in *73* magazine.[3] Wanting to replace the portable rhombic I'd been using for local and regional DXpeditions, I decided to use ZL4TAH's rope-support idea in building a very long 2-meter antenna based as much as possible on N6NB's long-boom quagi design. The result is an antenna that's designed for temporary portable operation, but one that can be used in more permanent installations with slight changes.

Design and Construction

The first eight elements (starting with the relector, through director 6) of this antenna are right out of any recent *ARRL Antenna Book*[4] for the long-boom quagi 2-meter antenna. The current *ARRL Antenna Book* includes two 2-meter quagis, one for 144.5 MHz and another for 147 MHz. Director elements 7 through 33 are what I added to this design. All added directors are the same length—$1/16$ inch shorter than director 6 (35 inches)—making directors 7 and higher $34^7/_8$ inches long for the 144.5-MHz version. The spacing between directors 6 and 7, and between higher directors, is also the same: $31^1/_4$ inches.

If you want to make a 147-MHz version, use the *ARRL Antenna Book* design, but again make directors 7 and higher $1/16$ inch shorter than director 6. (The spacing of $31^1/_4$ inches for directors 7 and higher should still be okay.)

The rope I used was $1/4$-inch yellow polypropylene that an electrician friend gave me. Although it's strong and lightweight, it's not a good choice for a permanent installation because the ultraviolet energy in sunlight will destroy it in a couple of months. (Although it's more expensive, Dacron would be a better choice for a long-term use because it's much more sun-resistant.)

I placed the antenna's two rope supports 26 inches apart, and centered the elements on them (Figure 1). My backyard lot is 100 feet wide, so I stretched the rope between both of my lot's fences to construct the antenna.

The directors consist of $1/8$-inch welding rod (aluminum fence wire and other similar materials will also work.) I tied them to the ropes with string and then used bathtub caulking compound to permanently attach each element in place. (I'm sure that other element mounting materials and methods could work just as well or better.)

The driven element and reflector frames (Figure 2), as well as the end support, are made from 1×2-inch lumber. The driven

Figure 1—Two runs of polypropylene rope support the antenna, here seen from its director end. Because the antenna exhibits something like 20 dB of gain (that is, it multiplies the power applied to it by a factor of 100), this is *not* the place to stand when the transmitter is operating!

Figure 2—1×2 lumber supports the antenna's reflector and driven elements, which consist of #12 TW house wire. It's not pretty—it just *works*.

The rope-ladder quagi's traveling weight suggests a friendly equation: 20 dBd of gain for 6 pounds of antenna!

How the directors mount on the ropes.

A look at the reflector end shows the reflector and driven element mounting details. (Note: The antenna was set up in this location for the photo. Don't aim it toward people or houses while transmitting.—*Ed.*)

element rope is terminated on the horizontal section of a 1×2 cross by a rope knot through a hole. Holes in the vertical section of the driven-element cross pass the #12 TW house wire that forms the element (you don't have to remove the insulation). Another set of holes, approximately $1/2$ inch from the director element rope support in the horizontal section, space the driven element from the reflector. The wooden cross section is therefore pulled from both directions—from the reflector side and the director side—with both ropes secured by knots through a hole in the cross section. At the center of the wooden cross, a $1/4$-20 bolt with a thumb nut allows scissor compaction and easy tear-down. A **U** bolt at each end provides a support mechanism.

Be sure to install the driven element's connector in a driven-element side that's in the same plane as the rest of the antenna. (Some quagis were made wrong by having the elements in the vertical plane but the connector section in the horizontal plane. This cross polarizes the driven element and directors!) Quagi builders' efforts to make absolutely square driven elements and reflectors is probably due more to aesthetics than electronics, but you may want to mark their bending points so you can easily reassemble the antenna after tear-down.

Another construction tip is that using silver solder when soldering the connector can give you a connection that's mechanically stronger than one made with 60-40 tin-lead solder. (Little rolls of silver solder that have their own internal flux are available at hardware stores for under $2.) Silver solder requires a little more soldering heat, but most soldering guns or irons should be adequate. It's a good investment for making any connection that's subject to vibration or needs extra strength.

Polarization and Feed

If you want your version of this antenna to be vertically polarized, turn the entire antenna 90° on its axis, and route the feed line backward through the reflector, and then down the back support. This minimizes interaction between the antenna and feed line to keep the antenna's polarization pure. (Feed-line routing isn't a problem if you configure the antenna for horizontal polarization because the coax comes straight down at right angles to the elements.)

Although a quagi's driven element is balanced, we traditionally feed it with coax, which is unbalanced. Although this is technically a no-no, N6NB found that a balun caused more loss than it was worth and eliminated it. Purists can add an RF current balun by winding the feed line into a 6-inch-diameter, 3 to 4-turn coil right at the connector, or by slipping 50 FB-73-2401 ferrite cores over the coax right at the connector. Like N6NB, I've found a balun to be unnecessary with my quagi.

Performance

With the antenna stretched across my front lawn, I measured its SWR as 1.4 at 144.5 MHz (SWR was lowest just below 144 MHz).

Testing the antenna at the West Coast VHF/UHF Conference in May 1986 revealed a gain of 20.6 dBd. This figure may or may not be correct because the antenna measuring team wasn't prepared to measure an antenna without a boom running down its center. Exact measurements notwithstanding, experience suggests that this antenna has considerable gain. Once, I used it with the "fox" transmitter in an all-day hidden transmitter hunt. I told the hunters ahead of time that the transmitter would use a horizontal antenna and operate at the low end of the band (both unusual in foxhunting) but I didn't tell them the reason: This 100-foot quagi is easier to mount horizontally than vertically, and was designed for 144.5 MHz! I pointed my 40-W signal at Mt Baldy (a great southern California bounce) from near Warner Springs. Some hunters went from the starting point (Palos Verdes) to Crestline and got out of the cars to hunt on foot *nearly 100 miles away* because the signal was so strong!

Additional Thoughts

With approximately 15 pounds of pull applied, this antenna droops 18 inches at its center. (Of course, you can reduce the droop—within limits—by pulling harder on its ends.) It's hard to say how much this droop affects the gain. Its near field is very large compared to 18 inches of droop, so my guess is there's little effect on gain. In any case, the droop can't be too bad because the antenna's effectiveness speaks for itself!

The antenna could be built for other bands using the same rope-ladder technique, but remember that the ends of half-wave elements are high-impedance points. You shouldn't use even high-quality insulators at the element ends; otherwise, detuning or gain reduction may occur. For the same reason, it's probably best not to make the distance between the support ropes greater than 75% of the elements' length.

Stacking two or more similar beam antennas—feeding them in tandem, not necessarily one above the other—is one way of getting more gain and a narrower beam-width. Some of the books I've read say that stacking two antennas efficiently can result in up to 2.7 dB of additional gain. But considering that the optimum stacking distance for Yagis is generally on the order of three-fourths of a single antenna's boom length—75 feet for this antenna—the feed line connecting both antennas had better be very low loss or the net stacking gain will be small. For example, graphs in my handbooks show 3.2 dB loss for 100 ft of RG-213 coax at 150 MHz. For 75 feet of the same cable at the same frequency, the loss would be 2.4 dB. Subtracting 2.4 dB from the maximum achievable stacking gain (2.7 dB) gives a net stacking gain of 0.3 dB—hardly worthwhile unless narrower beamwidth is what you're really after. With much lower loss coax, stacking might make some sense. Also, reducing the stacking distance might lower coax losses more than the departure from optimum spacing. Reduced stacking distances generally result in cleaner antenna patterns.

At least two other approaches might improve this antenna's gain. One would be to optimize its element lengths and spacing. I make no claim whatsoever that this antenna is optimized, because the Yagi design software I use can't deal with 35 elements! I learned some things when trying to optimize much smaller antennas that had already been well-designed with antenna-design software. Changing a director's length by $1/2$ inch at 2 meters often resulted in a calculated gain change of only 0.02 or 0.03 dB. The changes were more dramatic for impedance than gain. This tells me that automatic optimization in antenna-design software (assuming that you're working with a good design to start with) is a really neat feature. (The old adage, "Yagis *want* to work," is true. I hope that the elements I've added to N6NB's fine design prove it.)

The other way to increase this antenna's gain would be to make it even longer! I don't know when diminishing returns set in on a Yagi's length. (Even if diminishing returns come into play, making the antenna longer might still be better than stacking when you consider the wide spacing and coax losses involved in stacking two or more antennas like this.) If you have a lot of room, making the antenna longer might be the best approach to more gain. It's certainly not an expensive project!

Even if you make this antenna "just" 100 feet long, it's a good choice for communication over a difficult, fixed point-to-point path. I demonstrated its performance to a group of soon-to-be hams during the recent national testing day, which happened to fall in the moonbounce contest weekend. Despite our impromptu parking-lot setup, we heard moonbounce signals during one 30-minute stretch. These were probably superstations, but we weren't sure the antenna was pointed just right.

This is as much of an idea article as a construction project. What can *you* do with a lot of 2-meter antenna gain?

Notes

[1] Wayne Overbeck, K6YNB, "The VHF Quagi," *QST*, Apr 1977, pp 11-14.
[2] Wayne Overbeck, K6YNB/N6NB, "The Long-Boom Quagi," *QST*, Feb 1978, pp 20-21.
[3] A. H. C. Smirk, ZL4TAH, "The Sly Beam," *73*, Jun 1970, pp 74-79.
[4] Two-meter quagis also appear in the 1979 through 1989 editions of the ARRL *Handbook*.

By Keith Kunde, K8KK

From *QST*, February 1985

Try a "Dopplequad" Beam Antenna for 2 Meters

Loop antennas have taken many useful forms in Amateur Radio. Here is still another twist in loop-antenna design. Simplicity, low cost and good performance keynote the K8KK design.

The "Dopplequad," or twin-quad beam, is an interesting variation of quad antenna design that seems to have originated in Germany several years ago. German amateurs have done considerable development on this design.

I became aware of the twin-quad system through a description of it in The UHF Compendium, by Karl Weiner, DJ9HO.[1] The twin quad has attributes that make it an attractive choice for the VHF and UHF bands:

1) Horizontal polarization and relatively narrow vertical beamwidth help reduce noise pickup.

2) Fairly broad horizontal beam-width (over 60 degrees at the half-power points) allows coverage of a large sector while provoding the directional qualities of a beam antenna.

3) 9-10 dBd relative gain with good front-to-back ratio.

4) Simple feed requirements allow direct coaxial cable attachment with low SWR.

5) Compact, lightweight and easy-to-build design.

As a newcomer to the 2-meter band, I needed a general-purpose antenna that could also be used in an attempt to contact the Space Shuttle Columbia during the STS-9 mission. With its reasonably wide horizontal beamwidth and compact dimensions, the twin quad looked like just the ticket; I immediately began construction.

I will not attempt to explain the theory that went into the original design of the twin quad; that already has been done by others, and a summary of their work may be found in *The UHF Compendium*. I wrote this article for two reasons: (1) to show the construction details for amateurs who might want to build the antenna and (2) to encourage further experimentation and refine-

ment of the design by those with more time and equipment than I have. Here's how I built a twin quad for 2 meters, and some observations on its performance.

The twin quad consists of two vertically stacked quad loops connected in parallel. They are backed by a reflecting plane consisting of three or more horizontal rod-type reflectors. The driven and reflecting elements are supported at their center points by means of a rigid, lightweight frame.

The driven loops are a wavelength in circumference. Each side of a loop is a quarter wavelength at the design frequency. A single one-wavelength quad loop in free space has a radiation resistance at resonance of 100 to 140 ohms. In the twin quad, two loops are connected in parallel, so the radiation resistance should be about half that, or 50 to 70 ohms. In the "real world," the presence of the reflectors, their spacing from the loops, and other environmental factors, all have an influence on the effective radiation resistance of the antenna. More on this later.

When a quad is made of wire that is a tiny fraction of a wavelength in diameter (for example, a quad built for the HF bands), the sides must be made slightly longer than a quarter wavelength because of the lengthening effect caused by forming the wire into a square, as explained by Orr.[2] When building a quad for use at VHF and above, however, it is practical to make the quad loops from copper tubing or heavy gauge wire that is a much larger fraction of a wavelength in diameter. This seems to cancel out the lengthening effect so the sides can be made exactly a quarter wavelength long.

Calculate the quad loop dimensions by using the following simple formulas:

$$L_{(feet)} = 984 / f_{(MHz)} \; 1\lambda \qquad (Eq.\ 1)$$

$$L_{(inches)} = L_{(feet)} / 4) \times 12 \; \lambda / 4 \qquad (Eq.\ 2)$$

If you prefer to work with metric dimensions, the equivalent formulas are:

$$L_{(meters)} = 300 / f_{(MHz)} \; 1\lambda \qquad (Eq.\ 3)$$

$$L_{(cm)} = (L_{(m)} / 4) \times 100 \; \lambda / 4 \qquad (Eq.\ 4)$$

Using these formulas, I designed my antenna for resonance at 146 MHz, so each side is 20.22 inches long.[3]

The number of reflectors used isn't critical, but antenna gain will be improved somewhat with a larger number. In fact, a reflecting screen or plate could be used instead of the rods, but the slight increase in gain is probably not worth the added complexity in construction. This is especially true for a 2-meter antenna with a large screen. The Germans show designs using three or seven reflectors in The UHF Compendium (most of them have three). I decided to use five reflectors; to my eye, a twin quad with only three reflectors just doesn't look right no matter how well it works.

The reflectors are slightly greater than 0.5 wavelength long. Use the following formula to calculate the proper length (the value of L was calculated in Eq. 1):

Reflector length (in) =
  $[(L_{[ft]} \times 1.025) / 2] \times 12$ (Eq. 5)

For metric dimensions:
Reflector length (cm) =
  $[(L_{[m]} \times 1.025) / 2] \times 100$ (Eq. 6)

For 146 MHz, the reflectors are 41.45 in (105.3 cm) long.

The vertical spacing between reflectors is not critical. If an odd number of reflectors is used, one can be placed in the center of the rear frame, with the rest evenly spaced from the center one. Some sample reflector-to-reflector spacings (measured on centers) for 2-meter antennas are shown in Table 1. The seven-reflector model built by DB8NP also incorporates director elements mounted in front of the quad loops; the spacing shown would require a longer frame than the three- or five-reflector designs.

The spacing of the driven elements from the reflectors can have a significant effect on the performance of this antenna. I wasn't able to derive a formula for predetermining the proper spacing during my limited experiments. The Germans used driven-element-to-reflector spacings of approximately 11 inches, but I got a lower SWR reading across the band at 12.5 inches. What effects the wider spacing might have on gain, front-to-back ratio and antenna radiation resistance are not known to me, but the overall performance of the antenna is satisfactory.

The Framework

The framework that supports the antenna elements must be strong enough to keep everything in place in a stiff wind. Otherwise, there is nothing critical about it. I think the best material to use in making the frame is 1 × 1 inch square aluminum tubing. You could also try making it of wood. The frame need only be long enough to support the quad loops and the desired number of reflectors. For 2 meters, a frame 62 inches long is about right. This length has the added benefit of allowing the front and rear booms, and both cross members, to be cut from two 6-foot lengths of aluminum stock. Experimentation will be easier if the front-to-back spacing can be adjusted easily. I temporarily substituted 1- × 1-inch wooden cross members of various lengths during the measurements, and replaced them with aluminum stock that was cut to

Table 1
2-Meter Antenna Reflector Spacings

Refl. No.	In	Cm	Builder
3	20.1	51	DL7KM
5	15.0	38	K8KK
7	11.8	30	DB8NP

the same length in the final assembly.

Your first task is to make the quad loops. To make the loops in the simplest way, you will need about 15 feet of very-heavy gauge, solid-copper wire. The Germans used some heavy grounding wire, 5 mmin diameter, which is about the same as no. 4 AWG. The wire can be insulated or bare, but solid-copper wire in this gauge is not available everywhere. I suggest that you try electrical supply houses Or scrap-metal dealers in your area.

If no. 4 wire can't be located, you can use 1/4-inch copper tubing. This is available almost everywhere, and sells for about 30 cents per foot. The problem with copper tubing is that it will kink if you bend it too sharply. This can be solved with the aid of a spring tubing bender, commonly available at hardware and plumbing stores. These economical tools (I bought one recently for $1) will allow you to make reasonably sharp corners, down to about 1 inch in radius, for 1/4-inch tubing. The spring bender is used by slipping it over the tubing so that it overlaps the point where you want the bend. The coils of the steel spring support the tubing walls while the bend is made, and then the bender is simply slipped off. If the bend isn't quite right, you can put the bender back on and make further adjustments.

Quarter-inch-diameter copper tubing is soft, with little structural strength, but it should serve well enough at 2 meters and above. The rigidity of the tubing can be improved, however, if you can locate some no. 6 AWG wire. This wire, alone is a little too light for 2-meter quad loops, but it is the right diameter to slip down the center of 1/4-inch copper tubing. A light coating of grease or Vaseline spread on the wire will make this easier; the wire should be bare. If you follow this route, make each quad loop individually and join them later.

With wire-cored tubing, the spring bender is not needed because the wire will prevent the tubing from kinking. But a bending jig of some sort will help you construct square ioops. A bending jig can be made of 3/4- × 3/4-inch wood strips that are spaced exactly 1/4 inch apart and screwed to a piece of plywood. One strip should have screws only at the ends, so that a C clamp can pinch the strips together in the middle to keep the tubing from slipping. A short piece of 3/4-inch wooden dowel, screwed down at one end of the strips, will help you get a smooth bend in the wirecored tubing. After bending the tubing around the dowel

by hand, place a block of wood along the tubing (after the point of the bend) and tap it with a hammer in order to get sharp corners. Another set of wood strips, mounted at 90° to the first, will help prevent shifting as each bend is made. Make the jig so that the loops conform to dimensions you calculated earlier, as measured to the outermost edges of the loops.

When all four sides have been formed, the ends of the tubing will have to be bent in the reverse direction at a 45° angle with spacing of about 3/4 inch center-to-center (Figure 1). The ends should then be cut off squarely, leaving stubs about 1/2 inch long. The two loops will be spliced together at these stubs later.

Loops made on a jig like this will turn out square and flat, and it doesn't take much time to throw a suitable jig together. Of course, the jig can also be used for solidwire loops; just change the spacing of the wood strips to fit the wire diameter.

An alternative is to make the quad loops from lighter-gauge wire and to support the outer corners with some kind of crossarm. Be aware that in the twin-quad design the outer corners of the loops are the points of highest voltage, whereas the corners along the boom are the current points. Thus, insulation requirements along the boom are not too critical, but you need good insulation at the ends of the crossarms to minimize losses. You might consider making the entire crossarm from a good grade of insulating plastic or fiberglass. Also, the loops might have to be made slightly larger than one wavelength when using smaller wire, because of the lengthening effect mentioned previously.

Next, make the frame that will support the antenna elements. I made my frame and reflectors entirely from Reynolds aluminum extrusions (Figure 2). They are available in many hardware stores, but you can make the frame from wood if your budget won't support the more expensive aluminum. Here's a list of Reynolds extrusions you will need:

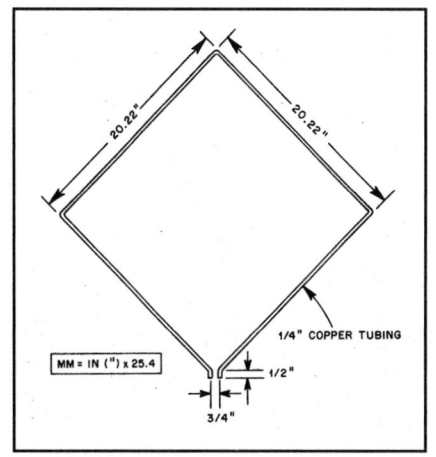

Figure 1—Dimensional details for one of the loop elements for the twin-quad 2-meter antenna.

Figure 2—Structural data for the aluminum frame used with twin-quad antenna (see text).

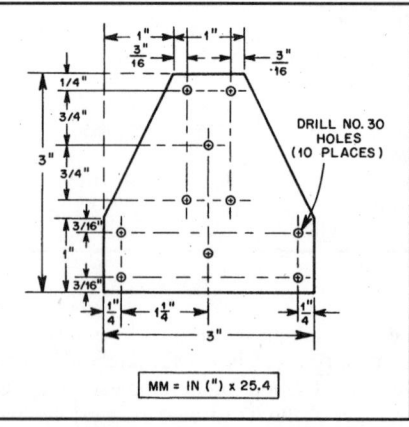

Figure 3—Drilling template for the antenna gusset plates.

Figure 4—Close-up view of the twin-quad antenna feed point.

• 2 — no. 4860 1-inch-square × 0.047-inch wall tubing, 6 feet long
• 2 — no. 1807 3/8-inch-diameter rod, 8 feet long
• 1 — no. 1806 3/8-inch-diameter rod, 6 feet long
• 1 — no. 2420 1 × 1 × 1/16-inch angle, 6 feet long

This list provides for five reflectors. You can omit one piece of no. 1807 stock if you are going to use only three. Also, the last item (angle stock) might not be needed, depending on how you want to mount the quad loops to the boom and the boom to the mast. This is discussed later.

While you are buying the aluminum, you might purchase an aluminum cookie sheet (the kind with no sides) to use for making the gusset plates. A radial-arm saw with a fine-tooth blade is excellent for cutting aluminum. A carbide-tipped blade is even better if you will be doing a lot of cutting. Otherwise, a hacksaw and a pair of snips will do the job. I cut both pieces of the square tubing to a 62-inch length. This left two pieces about 10 inches long to be used as cross members.

Cut the rod stock to length for the reflectors. If you should make a mistake in cutting these, make them a little too long! Drill (no. 29) and tap a hole for a no. 8-32 screw at the center point of each rod. Use plenty of oil while tapping, and remove the tap at the halfway point and brush off the chips. A little more oil, and the job is done.

Cut the cookie sheet into eight 3- × 3-inch plates. Then trim off two corners as shown in Figure 3 (use tin snips to trim the corners). I decided to use 1/8-inch-diameter × 1/8-inch-grip-range Pop Rivets to fasten everything together. The rivets are available in aluminum and won't rust in contact with the aluminum frame. Center punch and drill one gusset plate with a no. 30 drill. Then use it as a pattern to mark, center punch and drill all eight plates. Unless you plan to make "tons" of these things, don't waste time on them. Deburr the holes with a larger drill, going in just far enough to remove the burr.

The placement of the reflectors should now be marked on the rear boom, and 3/8-inch holes drilled completely through from one side. A drill press will help you make these holes square to the boom. Turn the boom 90° and drill clearance holes for the reflector locking screw. You are now ready to mount the antenna to the mast. Just clamp it to the mast with a couple of long U bolts. The mounting bracket assemblies I used can be omitted (these were made from 1-inch angle stock and 1/8-inch-thick aluminum cut from a 3 1/2-inch rack panel). If you decide to skip the brackets, you can now mark and drill the holes for the gusset plates and rivet them in place. Insert the two outermost reflector rods and lock them in with no. 8-32 × 3/4-inch stainless-steel screws with lockwashers. The remaining reflectors will be put in after the two booms have been linked together.

More Assembly Data

The front boom can now be prepared. Mark and drill the holes for the gusset plates, but do not rivet them in place yet. A satisfactory method of mounting the quad loops to the boom must now be found. The Germans show a sturdy assembly, that has a grooved plastic block with a plastic cover plate that clamps the loops to the boom. These blocks can be built up in layers from 1/4-inch Plexiglas if you don't have a source of thicker plastic. This is the method I recommend. But, I had a number of nice ceramic insulators in the junk box (3/4 inch square by 1 inch high), so I used them. Always put a cushioning gasket between insulators and their mounting surfaces if there will be any significant weight or stress applied to them. Thin cork sheeting, available from the plumbing department of your hardware store, is an ideal material. The quad loops are fastened to the insulators with brass clamps made of 3/4-inch-wide brass-strip stock, such as may be found in hobby shops. The two middle insulators are mounted on brackets of 1-inch angle stock that are riveted to the boom.

The final placement of the insulators should wait until the brass clamps have been made and test-fitted to the loops. Then, bolt everything together, square up the loops to your satisfaction, and solder all four clamps to the loops. If desired, the loops may be joined beforehand by soldering a sleeve of 1/4-inch-ID brass tubing over the end stubs. This makes it easier to handle the loops. The resulting assembly is sturdy, but the plastic-block clamps should be even better: They eliminate the need for the additional brackets, insulators and brass clamps.

Final Assembly

At this point, the antenna is ready for final assembly. Only the cross members linking the two booms have to be cut, drilled and riveted in place. This is where you may want to experiment. As stated earlier, the Germans used a loop-to-reflector spacing of about 11 inches. I used 12.5 inches, as measured on centers. Changes in the spacing have an effect on the SWR. If you aren't in an experimenting mood, use the leftover pieces of square tubing, uncut, as cross members. This will give you a spacing close to the 12.5 inches I found to be optimum after trying a number of wooden cross members of various lengths. The total spacing depends somewhat on the height of your loop mounts. Insert the remaining reflectors and lock them in place.

The parallel loops of the twin quad present a close enough match to 50- or 75-ohm line so that the line may be connected directly to the center point of the two loops (Figure 4). The shield should go to one side and the center conductor to the other. This will undoubtedly cause a few raised eyebrows, as it is well known that a balanced antenna (like the twin quad) should be fed with a balanced feed system. An unbalanced feed can cause disruptive effects, as documented by Orr, and by Maxwell in his article about baluns.[4,5] I tried feeding the twin quad with a quarter-wavelength coaxial-sleeve balun transformer, but could not get a satisfactory SWR at any loop-reflector spacing. The balun was properly resonant. It showed low SWR when connected to a dummy load. Another experiment involved placing a number of large ferrite beads over the outer conductor of

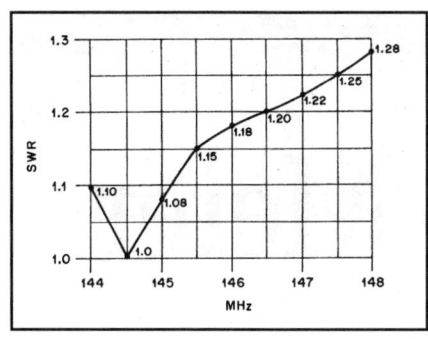

Figure 5—SWR curve for the completed antenna.

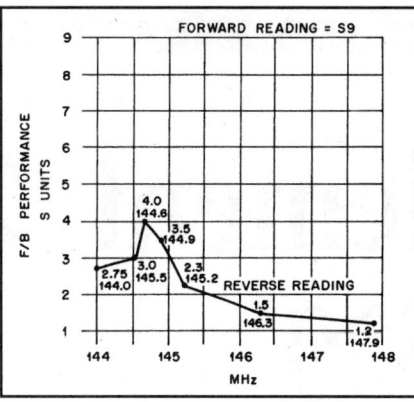

Figure 6—Relative front-to-back ratio of the twin-quad beam antenna in S units. (The actual value of dB/S unit of the author's Yaesu FT-726 transceiver is unknown.)

the coaxial cable near the feed point. If there was a lot of RF on the outer shield surface, the beads would choke it off. But again, the system SWR was adversely affected, so I returned to using a direct 50-ohm coaxial feeder.

The Germans also use the direct-feed technique, but have made measurements that show a small shift in directivity (about 6°) occurs with unbalanced cable. I also noted this, with the shift being in the direction of the side where the cable center conductor was connected. This small shift is nothing to be concerned about. But, what about other adverse effects?

There is no doubt that the length of the feed line affects the SWR readings of this antenna. However, I'm not so sure that the potential for radiation from the line impairs antenna performance. For instance, the directivity of the antenna seems to be good, and the Germans experienced little impact on the beam pattern from the direct coaxial cable feed.

One phenomenon you should be aware of is that you cannot make SWR measurements accurately if your SWR meter is too close to the antenna. This may be caused by the effects of RF on the outer shield of the feed line near the antenna. These effects gradually disappear a few wavelengths away from the antenna. The total feed line during my tests was 36 feet 9 inches of new RG-8/U. Measured line loss was about 1 dB at 146 MHz. No "hot-shack" effects were noted at the transmitter end of the line. Do not ground any part of the coaxial line at the antenna end. The cable should be brought straight back from the feed point, then down behind the reflectors. (The photographs were taken during tests, before final connection of the feed line.) Of course, the transmitter end should be well grounded. I suggest using a gamma or T match with the twin quad, but admit that feeding of this antenna definitely needs further exploration.

The SWR curve is shown in Figure 5. It appears that the resonant point is closer to 144.5 MHz than the intended 146 MHz, but the SWR is within acceptable limits across the entire band. (If I were to make another twin quad, I would make the loops about 1% smaller.)

Front-to-back measurements were made by aiming the antenna at a low-power oscillator placed on top of a stepladder in a field about 200 feet away. The receiver RF gain was adjusted to show an S-9 reading. Then the antenna was rotated 180° and another reading was taken off the back. Finally, the antenna was rotated back to the forward position to confirm that the S meter still read S 9. The plot in Figure 6 shows the front-to-back performance across the band. Rejection of signals arriving from the rear is quite good for an antenna of this simplicity.

Conclusions

I will leave forward-gain measurements to those with the tools and time to make them properly, but I have no reason to doubt the 9-10 dBd that the Germans claim. The antenna performs well, although I did not succeed in contacting the Shuttle. I was able to hear the operator several times, however, and the signals were loud.

The noise level is indeed lower with the twin quad than with a vertical. An ordinary TV antenna rotor will easily handle the twin quad.

If you're looking for a manageable antenna project, the twin quad may be a good choice. It lends itself to a variety of alternative construction techniques, and gives good performance for your investment in time and money. Please see the reference literature for more information on this interesting antenna.

Notes
[1] K. Weiner, *The UHF Compendium*, available from Ham Radio's Bookstore, Greenville, NH 03048.
[2] W. Orr, *All About Cubical Quad Antennas*, available from W6SAI.
[3] mm = in x 25.4; m = ft x 0.3048.
[4] See note 2.
[5] W. Maxwell, "Some Aspects of the Balun Problem," *QST*, March 1983, p. 38.

By Ron Hege, K3PF

From *QST*, July 1999

A Five-Element, 2-Meter Yagi for $20

This antenna is easy on your wallet and easy to build!

In a matter of a few hours, you can easily build a broadband, 2-meter Yagi—complete with mounting hardware—for $20. The antenna offers a gain of about 10 dB, is lightweight, mechanically strong and rivals the performance of similar commercial antennas.

The antenna's low cost is made possible by modifying a RadioShack FM broadcast receiving antenna (RS 15-2163). For $19.99, plus tax, you get a 70-inch-long by 1-inch-square boom, a set of six $^3/_8$-inch-diameter elements, antenna-mounting hardware and two plastic end caps to seal the boom ends. In addition to RadioShack's antenna, you'll also need some nuts and bolts to remount elements, an 11-inch length of RG-8 (or similar) coax, an SO-239 connector and a 9×1-inch-long aluminum strip. This strip is cut into two pieces to fabricate a strap for the gamma match and a mount for the SO-239 connector. The thickness of the strip is not important as long as it can be bent easily and is strong enough to hold the SO-239 connector firmly in place. To close any unused holes and the tips of the elements, you'll need some noncorrosive sealant, such as RTV. Most amateurs I know have these items on hand. If you don't, you'll spend a few more dollars.

Element Relocation

Refer to the accompanying photo and Figures 1 through 3. First, open all the antenna elements to their fully extended positions. Three of the elements are attached to plastic insulators and are tied together electrically with stiff, crossed, bare-aluminum wires. Each of these three elements looks like a dipole broken in the middle at the plastic insulators. One element measures about 58 inches from end to end, another about 56 inches and the third about 43 inches. You'll not need the 43-inch element.

Cut the wires next to the rivets on the 43-inch element. Drill out the rivet holding the element to the boom and discard the element. Use a screwdriver and pliers to release one wire from beneath one of the rivets on the 58-inch element. Try not to damage the rivet. Pull the wire out and away from the rivet. Go to the remaining wire on the 58-inch element; its opposite end attaches to another rivet on the 56-inch element. Unwind the end of the wire from beneath the rivet on the 56-inch element and pull it towards the 58-inch element. You now have a single wire on the 58-inch element with one loose end. Pull that wire straight across to the opposite rivet that no longer has a wire under it. Use pliers and whatever force is necessary to loop the wire around and under the rivet head as was the original wire. Seat the wire *fully* beneath the rivet head (see Figures 2 and 3). I was able to get the wire fully seated by pulling hard on the wire with my hand and squeezing the wire under the rivet head using the jaws of Vise Grip pliers. If you cannot get the wire fully wound and seated under the rivet, drill out the rivet and replace it with a bolt and nuts. *Do not* cut the wire off at the rivet. Pull the wire back toward the opposite rivet and cut it off leaving a pigtail about $1^1/_2$ inches long. You may want to reseat the rivet by hitting it with a hammer. Just be sure to back up the rivet's head with a hard object before striking the rivet's opposite end. Be careful not to damage the plastic insulator. You have now turned a two-piece element into a one-piece element, and this will be the driven element.

Drill a mounting hole in the boom (for the one-piece element) 17 inches away from the center of the adjacent 66-inch element (reflector). Remove the 58-inch element from its original location and mount it at the new position using a bolt, two washers and a nut. Place one washer directly against the plastic insulator under the wire that connects the two 3/8-inch-diameter tubing halves together. Position the other washer on top of the wire so it bears down on the wire when the bolt is tightened. This puts the center of the element at the same electrical potential as the boom. Using the $1^1/_2$-inch pigtail, bend it and place it between the two washers so there is a piece of wire on each side of the bolt. This prevents the washers from tilting and makes for a cleaner fit. Trim off any excess wire. (All of the foregoing is more difficult to describe than it is to perform! It doesn't take long to do once you understand what is going on.)

The next element (56 inches long; Director 1) is handled similarly to the preceding one. However, this element originally had *two* wires beneath each rivet head. One of those wires has already been removed. At the opposite rivet, unwind one of the two wires so that only one wire remains beneath each rivet. Pull one loose end of a wire straight across to the opposite rivet and force the wire into place under

2-16 Chapter 2

the rivet just as before. Pull the other loose wire end to its opposite rivet and force it into place. The two element halves should now be connected together with two wires. The wires will be parallel to each other and on opposite sides of the rivet that secures the element to the boom.

Next, drill a hole in the boom 13 inches from the center of the 58-inch element (DE). Remove the 56-inch element from (D1) its original location and mount it on the boom at the new hole. Again, place a washer on opposite sides of the wires so that the washers squeeze against the wires as the bolt and nut tighten the element to the boom.

The remaining three elements (REF, D2 and D3) don't need to be modified; their individual dipole sections are already joined by metal plates. All you need to do is remove two of them from the boom, drill new mounting holes and mount them at their new locations. The first 50-inch element (D2) is placed 16 inches—(center to center)—from the adjacent 56-inch element. The end element (D3)—also 50 inches long—is placed 21 inches (center to center) from the new location of its adjacent 50-inch element. All of the elements are now in place ready to be cut to length for 2-meter operation.[1,2]

Element Trimming

For this job, a fine-toothed saw works well. (Caution: During the following steps, be sure that you cut *half the total amount* from each half [ie, each side] of an element.) For operation at the low end of the band (144 MHz), cut the 66-inch element to a total length of 41 inches (see Figure 1B). This element becomes the reflector. Cut the next element in line (the driven element) to a length of $38^7/_8$ inches. Cut the next three elements (directors D1, D2 and D3) to lengths of $38^1/_8$, 37 and $36^5/_8$ inches, respectively. If you want to trim the elements for use at higher frequencies, cut $^1/_4$ inch off of each element for each 1-MHz frequency increase. For instance, cutting a total of $^1/_2$ inch from each element tunes and maximizes the antenna for 146 MHz. (Again, cut *half* the total amount from each half of an element section. For 146 MHz, the preceding example, that's $^1/_4$ inch from each half-element section.)

SO-239 Connector and Mount

Refer to Figures 4 and 5. Attach the SO-239 connector to the bottom of the boom beneath the driven element using an L-shaped piece of aluminum. Fabricate the bracket from a $3^3/_8$ inch length of aluminum cut from the 9×1 inch strip. Bend it at a right angle so that one side is about $1^1/_4$ inches long. Make the necessary holes to mount an SO-239 connector on the $1^1/_4$ inch long section and secure the connector to it. Fasten the bracket to the boom bottom using bolts and nuts, positioning the bracket so that the tip of the SO-239 center pin faces the reflector. Position the tip of the pin about $^3/_{16}$ to $^1/_4$ inch in front of the center

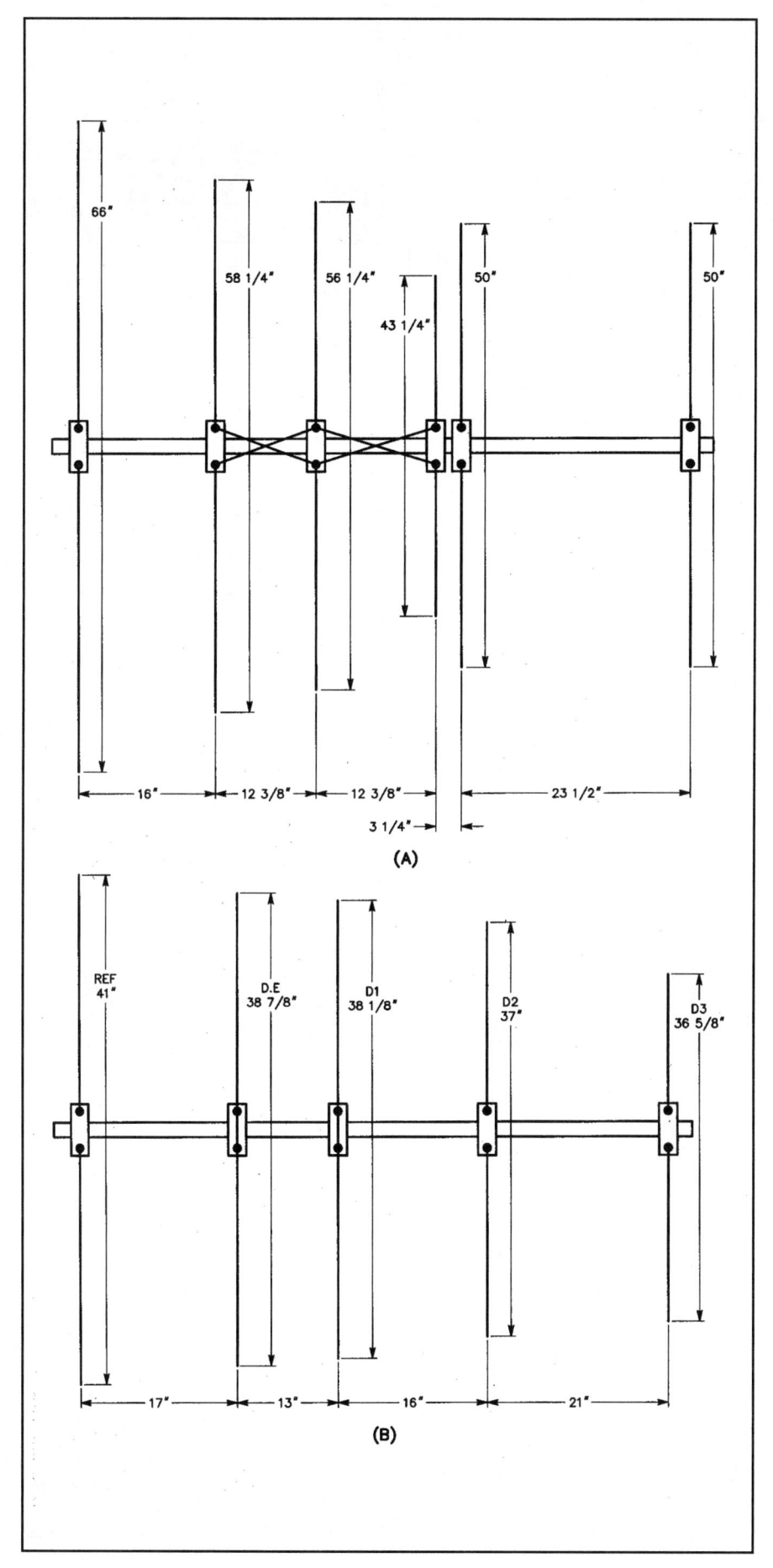

Figure 1—At A, the original configuration of the RadioShack FM receiving antenna. The element lengths and spacings at B are chosen for operation on 144 MHz. For operation on higher frequencies, shorten the elements even more; see text.

144 MHz 2-17

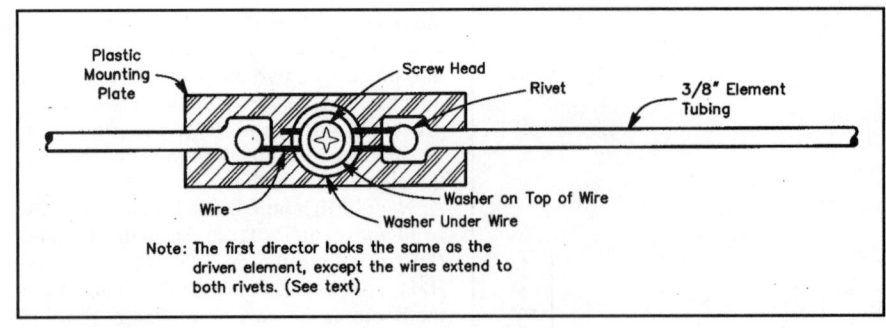

Figure 2—Drawing of the driven-element modification.

Figure 3—Here's the modified driven-element.

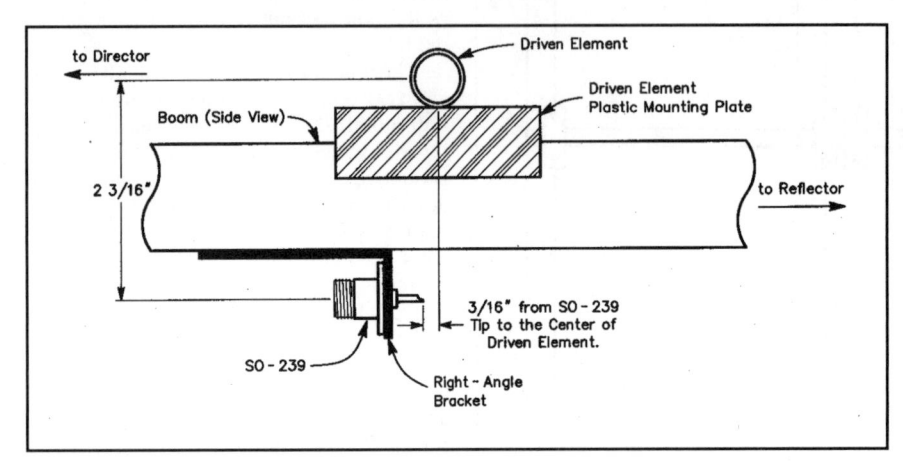

Figure 4—Side view of the driven-element area and SO-239 mounting bracket.

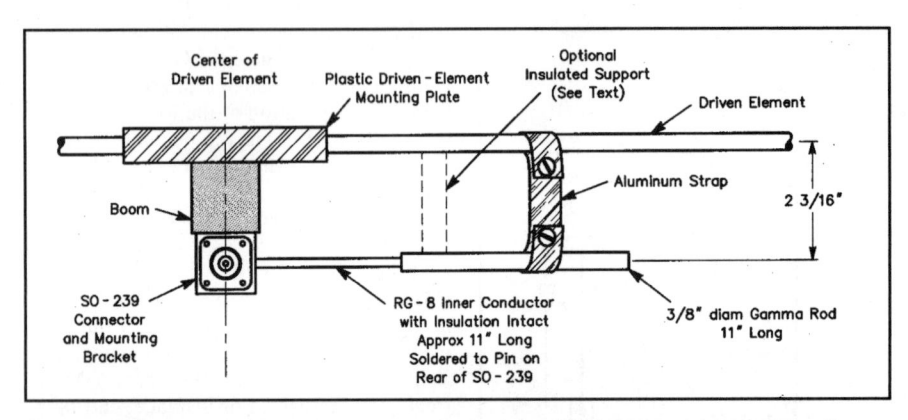

Figure 5—The 2-meter Yagi's gamma match. A piece of RG-8 coax and a length of tubing combine to create an inexpensive and rugged gamma-match capacitor.

of the driven element toward the director side.

Making the Gamma Match

Remove the outer insulation and braid from an 11-inch piece of RG-8 coax, leaving the center conductor and its insulation. Strip off 1/2 inch of the insulation and solder the center conductor to the SO-239 pin. At the pin, bend the wire at a right angle so that the wire is parallel to and about 2 3/16 inches away from the driven element along its length (see Figure 5). This lead forms the inner plate of the gamma capacitor. Next, select a piece of the scrap 3/8-inch tubing you cut from one of the antenna elements and cut it to a length of 11 inches. Slip this tubing over the RG-8 inner-conductor insulation to form the outer plate of the gamma capacitor. Position the tube seam so it faces the ground when the antenna is at its operating position; this allows moisture an easy way out. To complete the capacitor construction, wrap the remainder of the 1-inch aluminum strip around the driven element on one side and around the 11-inch tube on the other. Construct the strap so that the centers of the tubing sections are approximately 2 3/16 inches apart. Leave a tang on each side of the strap to accept a locking screw. Trim away any excess material.

Tuning the Gamma Match

Before applying RF to the antenna, connect an SWR meter to the SO-239 connector *at the antenna*, not at the transmitter end of your transmission line. This ensures that you are tuning *just the antenna*. For a quick adjustment of the matching network, you can try positioning the antenna straight up toward the sky, with the reflector sitting on the ground. Using this approach, however, I found that when I raised the antenna to a height of 10 feet on a metal mast, the gamma capacitor needed readjustment. If you're a perfectionist, it might ultimately be less work to tune the antenna while it's mounted in the clear a few wavelengths above ground or sitting at its intended operating position. If you're going to use a metal support mast, attach it to the antenna prior to tuning. Use a nonmetallic mast (wood, fiberglass, etc) if you're going to mount the Yagi vertically (so that the elements are in line with the mast); otherwise, antenna performance will suffer a bit. It's okay to use a metal mast when using horizontal polarization.

Reduce your transmitter's output power to about 1 or 2 W for safety use or, an antenna analyzer. Don't use more than a few watts—you don't need it. Set the transmitter frequency to that for which you cut the antenna. (Remember to ID your station during this adjustment period.) First, adjust the gamma strap (sliding it back and forth) on the driven element for the lowest SWR. Then slide the gamma tube (capacitor) back and forth within the strap for lowest SWR reading. You should be able to get a match by alternately making adjustments to the strap and gamma tube. I was able to tune my antenna to a 1:1 match. (An SWR of 1.5:1 or less is acceptable.) Recheck the SWR reading after finally tightening the strap to be certain that everything is still okay. Check by eye to ensure the gamma-capacitor tube is parallel with the driven element from one end to the other. It doesn't matter if the gamma-capacitor tube is slightly in front of or behind the driven element, but it should be parallel to it.

Performance

I don't have the proper equipment for making antenna-gain measurements. However, I made a crude comparison of the Yagi to a dipole using the following approach: First, I erected a 2-meter dipole on a 10-foot-long metal mast and adjusted the antenna for a 1:1 SWR. While feeding the dipole with a few milliwatts, I placed my H-T about 75 feet away from the dipole. The S meter reading on the H-T went full

Figure 6—The gamma match and driven element.

scale, so I removed the antenna from the H-T. The reading was *still* full scale. I then wrapped the H-T case with a shield of aluminum foil and the S meter reading dropped to S3. After that, I never touched or moved the H-T throughout the rest of the test. I rotated the dipole 90 degrees and, as expected, I got a zero reading on the H-T's S meter. I then turned the dipole back to its original position and rechecked the S meter. Again, it read S3. I used this reading as the dipole reference. Then, removing he dipole, I replaced it with the Yagi. I pointed the Yagi directly at the H-T and fed it with the same power level used to feed the dipole. The S meter read full scale on my H-T! Because S meters are notoriously inaccurate and not calibrated, I have no way of knowing how much gain that indicates, but it's a lot! (A five-element Yagi on a boom this long is capable of producing a gain of about 10 dB.[2]) When I turned the Yagi so that its reflector faced the H-T, the meter reading dropped to S4. That's a nice front-to-back ratio![3] I also checked radiation off the sides of the antenna. I was pleased to see an S0 meter reading from each side. [*The ARRL Lab modeled Ron's Yagi using* YO *software and verified his claims.*—Ed]

Summary and Acknowledgment

After I finished my project, I decided that it might be a good idea to make sure that the gamma match worked okay when the elements were cut for 146 MHz because many readers might want to use the antenna for working distant FM repeaters. I sawed off $^1/_2$ inch from each element and went through the tuning procedure again. I was still able to get a 1:1 match. Those wishing to use the antenna for FM repeater work should orient the antenna elements vertically.

I wondered what effect moisture would have on the gamma capacitor. So, I poured water into one end of the gamma tube until it came out the other end. I rechecked the SWR and I found only a barely noticeable effect. I recommend you plug the ends of the tube with a dab of RTV or other noncorrosive sealant to keep out dirt and insects.

The driven element holding the gamma match will not fold for portable use if the bracket holding the SO-239 connector is bolted to the boom. One simple solution is to remove the bolts holding the bracket. Removal and replacement is made easier if you use wing nuts on the bolts. An optional, second support for the gamma-capacitor tube, made from nonmetallic material, provides better support for the tube during transportation.

My thanks to Larry, K3PEG, for instructing me about this type of gamma match fabrication.

If you're looking for a good 2-meter antenna, try this one! It's inexpensive, easy to tune and is the simplest construction approach I've seen for quickly "homebrewing" a 2-meter Yagi.

Notes

[1]The element lengths and spacing dimensions for this antenna are taken from page 631 of The 1974 *ARRL Handbook*.
[2]See also Edward P. Tilton, W1HDQ, *The Radio Amateur's VHF Manual* (Newington: ARRL, 1972), third edition, p 155, Figure 8-4.
[3]Using *YA* and assuming $^1/_4$-inch-diameter elements with no tapering, modeling the 1974 *Handbook* antenna shows a gain of about 10 dBi and a 9-dB F/B ratio.–*Zack Lau, W1RF*

By Harold "Hal" Thomas, K6GWN

From *QST*, January 1998

A 2-Meter Phased-Array Antenna

Eighteen months ago a newly licensed Technician dropped by and asked for some advice on VHF antennas. My friend had limited funds, so his first creation was a simple 2-meter vertical installed atop an old 40-foot tower. He used this antenna to work 2-meter SSB with some success, despite the fact that his antenna was vertically polarized. (Most SSB enthusiasts use horizontal polarization. The signal loss caused by polarization "mismatch" can be substantial.) During his SSB explorations he routinely heard operators working stations up the coast over distances of 200 miles and more. He also heard them talking about "Yagis" and "quads."

Seeking to better his station, he visited again and asked me to explain antenna gain, directivity and polarization. Firing up the computer and my *EZNEC* software, we sat down together and designed a horizontally polarized, 12-element Yagi for 2 meters with a 17-foot boom. I wasn't sure if he was up to the task of building such a thing, but I was wrong.

Phaseuth Arun, AC6NX, arrived in the States 10 years ago from Cambodia with no trade skills. While doing odd jobs he read voraciously and, in the process, became a self-taught computer scientist. (He is now a system programmer and network specialist for a major corporation.) At the same time, Phaseuth was bitten by the ham bug. He jumped feet-first into the hobby, allowing nothing to stand in his way. Six months after I first met Phaseuth, he upgraded from Technician to Amateur Extra. So, I wasn't surprised when I heard that he had assembled and installed the home-brew Yagi on his tower. Shortly thereafter, AC6NX worked Hawaii with 150 W!

Basking in the glow of Phaseuth's enthusiasm, I thought it was time for *me* to do something. Two-meter SSB didn't interest me, but 2-meter FM did. I could have simply erected a vertically polarized Yagi for FM work, but I wanted to do something different—just for the pure fun of it. Many hams use ground plane antennas for omnidirectional coverage on VHF, but how many have ever tried two ground planes in a *phased array* to focus the radiation pattern in a particular direction? This phasing trick works well for verticals on the HF bands, but what about 2 meters? I decided to try!

Ask the Computer

I modeled my idea on the PC and found that if I fed the two ground planes 90° out of phase with each other and separated by a quarter wavelength, I could theoretically achieve 4.75 dBi of gain and a front-to-

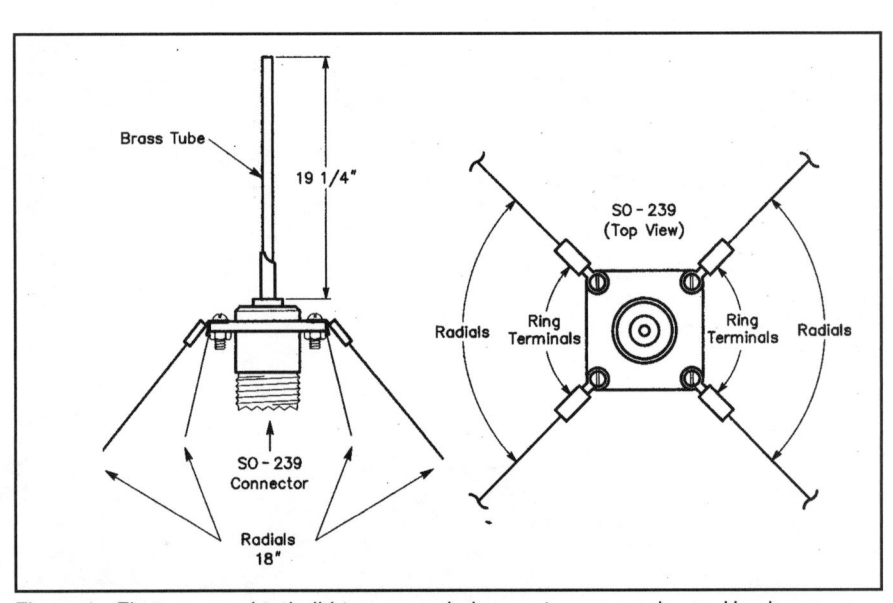

Figure 1—First you need to build two ground plane antennas as shown. Use brass tubes for the vertical radiators and stiff copper or aluminum wire for the radials. Ring terminals, screws and nuts secure the radials to the SO-239 chassis connectors.

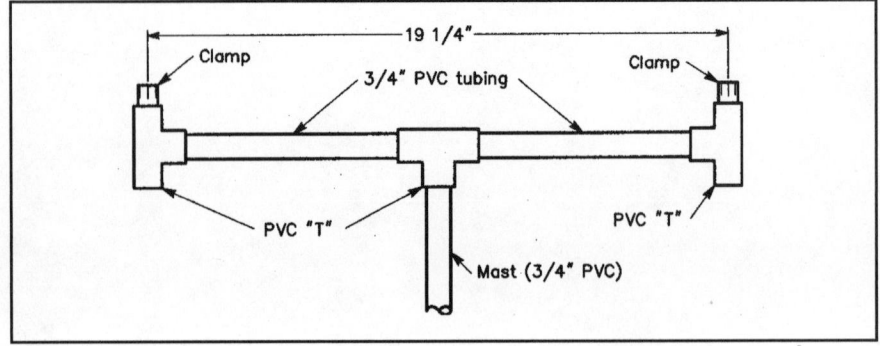

Figure 2—Assemble the boom using ³⁄₄-inch diameter PVC tubing and Ts. Note that the distance is 19¹⁄₄ inches between the center of one PVC "end T" and the other.

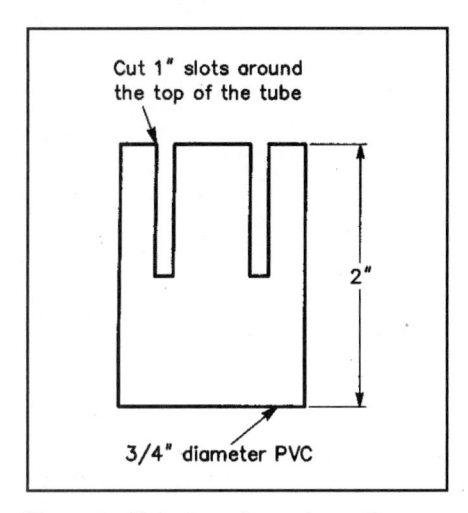

Figure 3—Make two clamps by cutting 1-inch slots into two 2-inch long, ³/₄-inch diameter PVC tubes. When you tighten the steel hose clamps around these slots, they'll apply firm pressure to the PL-259 connectors, holding the ground planes in place.

Figure 4—The phasing harness. The lengths of the individual phasing lines are critical and should be as precise as possible. The direction of radiation is toward the antenna connected to the *longest* line.

back ratio of about 40 dBi. That's an impressive amount of "focusing" for two ground planes, but could I do better? Sure enough, feeding the antennas 135° out of phase resulted in 6.29 dBi of gain, although the front-to-back ratio was somewhat reduced. I chose to go with the benefits of the 135° phase difference.

When I assembled and installed the phased array, I was pleasantly surprised to discover that the computer's predictions were very accurate. By aiming the antenna with a simple TV rotator, my effective range on 2-meter FM increased dramatically. When interfering signals became a problem, I was often able to turn the array slightly and reduce the offending signals a great deal. We're talking about taking a full-quieting FM signal and rendering it almost nonexistent!

The obvious question is, why build such a thing? Why not just use a Yagi on a rotator to achieve the same result? Not all hams have enough space for a 2-meter Yagi, especially if they intend to rotate it. My compact array is ideal for attic installations or any other environment where space is scarce. You get the benefits of a directional signal pattern—and the gain that comes with it—along with a small turning radius (less than 40 inches). Of course, building the array is also fun and educational!

Assembling the Ground Planes

Gather the parts you'll need to build the ground planes:

8 Ring terminals or long solder lugs
2 ³/₃₂-inch diameter brass tubes, 20 inches long
2 SO-239 chassis connectors
8 Radials, each 18 inches long (you can make these from stiff copper or aluminum wire)
8 Bolts, nuts and washers (4-40 × ³/₈ inch)

Solder the brass tubes into the SO-239 connectors (see Figure 1). Trim the tubes so that the total length from the flange of the SO-239 connector to the tip is 19¹/₄ inches. Using the 4-40 × ³/₈-inch screws, nuts and washers, install four ring terminals into each of the four holes on the SO-239 connectors. Solder the radial wires to the ring terminals. Bend the radials down at 45° angles on both antennas.

Now it's time to test and adjust each ground plane. Attach a piece of string or fishing line to the top of the brass tube (a little tape will do the trick) and suspend the antenna about six to eight feet above the floor or ground. Attach a length of coax (about 15 feet of RG-8X is fine) to the antenna and run it to a 2-meter transceiver and SWR meter. Measure the SWR at about 146 MHz and adjust for the lowest SWR by moving the radials up or down. If you've constructed the antennas properly, you should be able to adjust both for SWRs of less than 1.5:1.

Build the Boom

The boom is constructed from schedule-40 PVC tubing (see Figure 2). Take care to assemble it precisely. It's important to maintain the quarter-wavelength spacing between the ground planes.

You'll need:

2 Lengths of ³/₄-inch diameter PVC tubing 8¹/₄ inches long
3 ³/₄-inch PVC **T**s
2 Lengths of ³/₄-inch diameter PVC tubing 2 inches long
1 Tubing cutter (optional)
1 Can "purple primer" PVC cement
2 1-inch diameter stainless-steel hose clamps

Don't use standard PVC cement. This stuff hardens so fast that you won't have time to adjust the fit. Be sure to get purple primer or equivalent. It sets up in 12 to 15 seconds, plenty of time for adjustments.

Sand the ends of both 8¹/₄-inch tubes until they are smooth. "Dry fit" the entire boom by slipping on the vertical "end" **T**s and then sliding both tubes to the horizontal "center" **T**. Measure for a total length of 19¹/₄ inches between the centers of the end **T**s. Sand or file the tubes as necessary to achieve the correct overall length.

Now disassemble your dry-fit boom and clean all joints, inside and out. Put a tube into a bench vise and clamp. Swab cement on either end of the tube (your choice!) and on the inside of one of the end **T**s. Slide the center of the **T** onto the tube and quickly push it as far as it will go. Now do the same with the remaining 8¹/₄-inch tube and **T**.

Clamp the center **T** horizontally into the bench vise. Slide both tube/**T** assemblies into the open ends of the **T**. Rotate the individual tubes so that their end **T**s are properly aligned with each other. Using a marking pen, draw short, thick lines on the tubes where they meet the center **T**. Extend the lines a short distance onto the center **T** itself. Remove the tubes, prepare the tubes and center **T** with cement, and then reinsert. Quickly rotate the tubes so that the lines you just made with the marking pen match correctly.

Finally, we have to insert two small PVC sleeves that will serve as holding clamps for our ground planes. Cut two 2-inch-long pieces of your ³/₄-inch PVC tubing. Cut several 1-inch slots in the sleeves (see Figure 3). I used a hacksaw, then a wood saw to increase the width. Deburr the slots carefully. Slip the sleeves into the end **T**s and check for a proper fit. About one inch of each sleeve should extend above the **T**. (Make sure you install the sleeves in the *top* ends of each **T** with the slotted portions protruding!) Cement them into place and, when the cement is dry, slip on the stainless steel hose clamps—but don't tighten them yet.

This is a good time to cement a length of PVC tubing to the bottom of the center **T** so that you can attach the finished antenna to

144 MHz 2-21

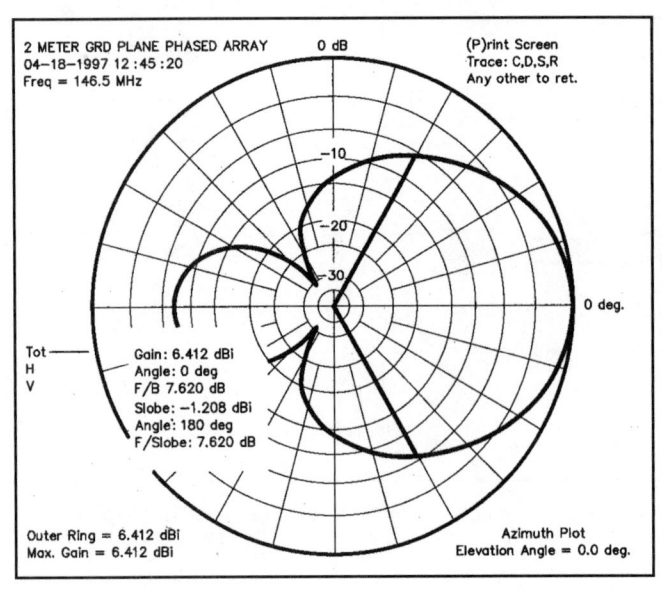

Figure 5—The theoretical radiation pattern of the 2-meter phased array, according to *EZNEC*.

Figure 6—The assembled phased array antenna. Note how the two phasing lines reach the ground planes through the bottoms of the PVC **T** sections. The entire array radiates in the direction of the ground plane that's attached to the *longest* phasing line.

a mast. I used a 2-foot length of ³⁄₄-inch PVC tubing.

The Phasing Harness

One way to feed two antennas out of phase is by varying the distance the RF energy must travel to reach each antenna. In our case, when the RF reaches the feed line **T** connector, it splits and travels on two separate paths to our ground planes (see Figure 4). This causes the RF to arrive at one ground plane well ahead of the other. As the ground planes radiate the energy, the phase differences that result from the different "arrival times" cause the signal to be canceled in some directions and reinforced in others. This creates the overall radiation pattern shown in Figure 5. Notice how it is focused in a particular direction. [The exact current amplitude and phase at each ground plane depends on the tuning of each element plus the exact velocity factor of the coax used. The change in gain is minor, but the depths of the side and back lobes may vary. The antenna will still work fine, however—*Ed.*]

So, let's get to work on our phasing harness. You'll need:

4 PL-259 coaxial connectors
1 UHF female **T** connector
1 Three feet of RG-8X coaxial cable
2 UG-176 adaptors

It is important to maintain the correct lengths of the phasing lines. Cut one length of RG-8X coax to 16 inches. Cut another to 22 inches. Solder PL-259 connectors on both cables, but take care not to shorten their lengths in the process. Connect one end of each phasing line to the UHF **T** connector.

Final Assembly and Testing

Route the unconnected ends of your phasing lines through the bottoms of the end Ts on the antenna boom (see Figure 6). Screw the PL-259 connectors into the bottoms of your ground planes. Pull the coax and PL-259s down through the **T**s until the antennas rest on the slotted sleeves. (You may need to rotate the antennas slightly to keep the radials from touching each other.) Tighten the hose clamps until they grip the PL-259s firmly. Use electrical tape to secure the coaxial **T** connector to the mast portion of your PVC tubing. Just allow the phasing lines to dangle in mid-air.

Install your antenna onto a mast or rotator, connect the main feed line to the bottom of the coaxial **T** connector and you're ready to go! The primary lobe of radiation will be in the direction of the ground plane that is attached to the longest cable of your phasing harness. Use your marking pen and draw an arrow on the boom to mark which direction your array is "pointing." This will be helpful when you install the antenna.

I installed my phased array on a rotator about 12 feet off the ground. After adjusting the radials a little for the best SWR (1.3:1 at 146 MHz) I was eager to get on the air. Using 300 mW I put a full-quieting signal into the Catalina Island repeater about 40 miles from my location. I increased my output to 30 W, turned the array, and worked a station *on simplex* in San Diego, 135 miles away! The longest contact so far was through a repeater near Paso Robles, about 200 miles distant.

I believe the 75° beamwidth and low angle of radiation are big factors in this array's performance. All it took was a $20 investment and about 3 hours of my time!

I would like to thank Phaseuth Arun, AC6NX, and Harry Amloian, KF6HBQ, for their help and encouragement.

By Nathan Loucks, WBØCMT

From *QST*, April 1993

7 dB for 7 Bucks

Need a 2-meter beam antenna, but you're short on cash? You can build this antenna for the cost of a fast-food meal.

When I became active on 2-meter FM, I soon discovered that I needed a beam antenna to hit some of the local repeaters. "Local" in Crosby, North Dakota, means 50 to 60 miles as the crow flies!

I knew there were plenty of VHF antenna projects in *QST* and other ARRL publications. Although many are easy to build, their matching systems are almost incomprehensible for beginners. Just take a close look at a typical matching network and try to figure it out. It's either some kind of a balun transformer (that you must wind yourself), a clampy, slidy stub match, or a nightmare of various coils and capacitors! I decided it was time to design, build and test my own beam antenna.

A Little Theory

My three-element beam is really nothing more than a half-wave dipole antenna (the driven element) mounted between two other elements known as the reflector and the director. Typically, the reflector element is about 5% longer than the driven element, and the director is 5% shorter. By spacing elements about 0.15 to 0.25 wavelengths from each other, you end up with a beam antenna with about 7 dB of gain. That is, the reflector and director act to shape the energy in a particular direction. By focusing the radio energy in this way, you concentrate the power in the direction you desire (just like a spotlight). That's gain!

To figure out the correct length of my driven element, I used the simple formula for a half-wave antenna above 30 MHz:

$$\text{Length (in feet)} = \frac{475}{\text{Frequency (in MHz)}}$$

At the feed point of a half-wave dipole, the impedance is about 72 ohms—close enough for an acceptable match for typical 50-ohm coaxial cable. Of course, we're talking about a beam antenna, not a simple dipole, so the impedance may be different. Armed with this basic information, I went to work!

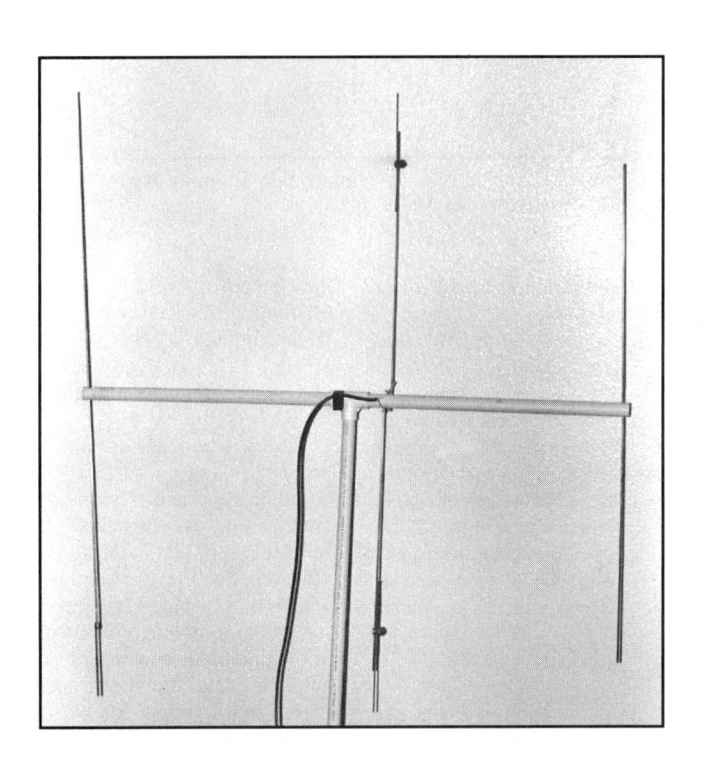

Construction

The construction is simple and cheap. I recommend $^3/_4$-inch PVC plumber's pipe for the supports. Two 18-inch pieces become the boom where the elements are installed. A 36-inch piece is used for the mast (see Figure 1). The boom and mast pieces are held together with a PVC T joint. Three small holes are drilled in the T as well as the boom and mast pieces. Insert the booms and mast into the T and rotate them until you align the holes. Now use screws to secure everything in place. You'll notice that my antenna is vertically polarized. This is best for FM work. If you want to use this antenna for 2-meter SSB and CW, just assemble the boom so the elements are parallel to the ground, not perpendicular.

You can make inexpensive elements from steel oxyacetylene welding rods. Hobby stores often carry brass rods that are suitable, too. Cut the director and reflector elements to the lengths shown in Figure 1. (If you can't find a 40-inch rod for the reflector, don't worry. You can solder small pieces at the ends to achieve the total length.) Drill holes in the boom about $^1/_4$ or $^1/_2$-inch from each end. I suggest you use a yardstick to draw a straight pencil line along the boom, marking an X for each hole. The idea is to keep the hole positions aligned as you drill them. Choose a hole size that offers a snug fit when you push the rods through.

After drilling, slide the director element through the holes and adjust it until you have an equal length on each side of the boom. Use solder or epoxy cement to hold it in place. Do the same with the reflector element.

Now for the driven element and matching network. But wait! This is too simple. It can't work! Well, my matching device uses a special gimmick. (Those of you from the old school know what I mean.) To con-

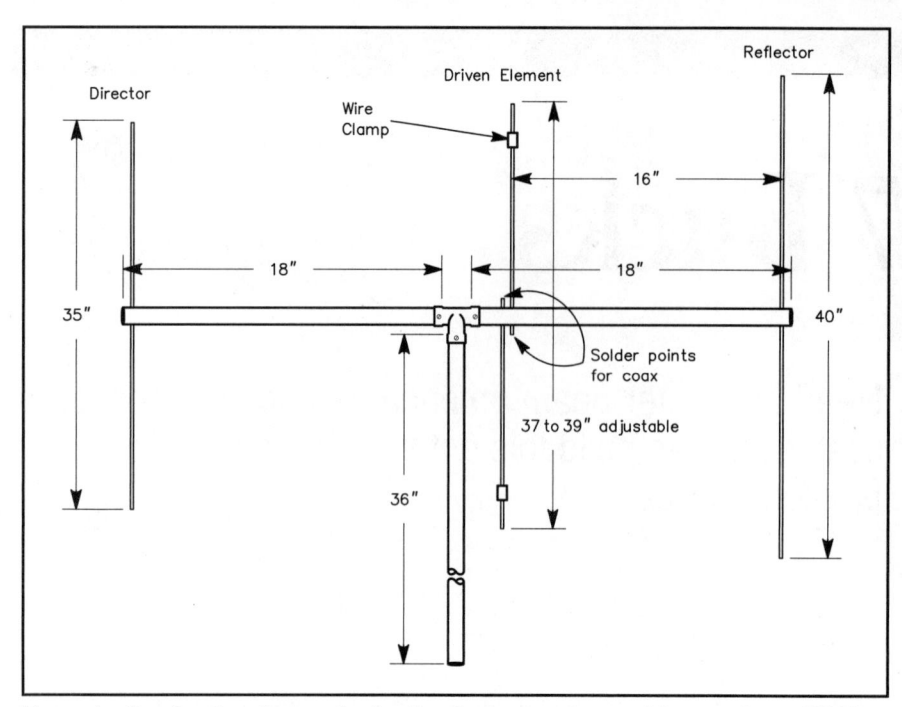

Figure 1—Construction diagram for the 2-meter beam antenna. A few sections of PVC pipe and a handful of welding rods are all you need.

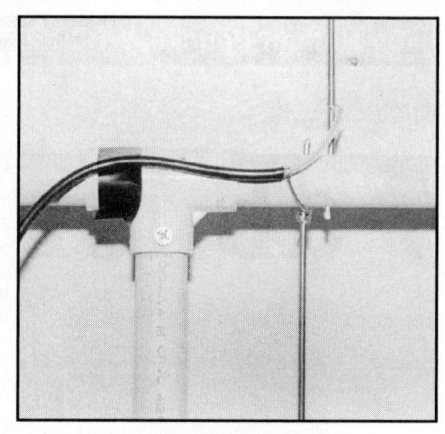

Figure 2—Solder the coaxial cable to the driven element stubs as shown. If the distance from your antenna to your radio is greater than 15 feet, use a low-loss cable such as RG-213.

struct the driven element, measure and mark the boom about 16 inches from the reflector. Drill two holes about $1/8$ to $1/4$ inch apart. Cut two rods, both 20 inches in length. Push the rods through the holes until $1/4$ inch protrudes from each side of the boom. You've just made the driven element and the matching device!

Solder your coaxial cable directly to the protruding stubs as shown in Figure 2. Use solder and/or epoxy to secure the rods to the boom. Apply electrical tape or use a silicon compound to weatherproof the solder joints.

Tuning

To make tuning the antenna easier over a wide range of frequencies, I clamped 4-inch pieces of welding rod to each driven element with #12 or #14 jam screws (electrical connectors). By loosening the screws and adjusting the rod sections equally, I can quickly tune the antenna for my favorite parts of the band.

Use a wooden step ladder to get your antenna at least five feet off the ground. Another approach is to tie a string to the antenna and use the nearest tree branch to hoist it 5 to 10 feet in the air. Once you have your antenna elevated and away from any nearby metal objects (such as aluminum siding and automobiles), place an SWR meter in the line between your transceiver and the antenna. Make sure to use an accurate SWR meter that's rated for operation on the 2-meter band. Now transmit and check the SWR reading. Adjust the clamp sections as necessary to get the lowest reading.

If you want to dedicate the antenna for operation on a relatively narrow set of frequencies (for a particular repeater, for example), don't use the clamp method. Just trim the driven element rods a small amount at a time until you get your lowest SWR reading.

After the antenna is tuned, mount it on the highest point possible (on the chimney or vent pipe, for example). Point the antenna in the direction of your favorite repeater, or use a TV antenna rotator to change direction remotely. If the distance from your radio to the antenna is more than 15 feet, use a low-loss coaxial cable such as RG-213.

That's all there is to it! It's simple, portable, tears down quickly, and best of all, it requires no complicated matching. What more could you ask of an antenna—for just seven bucks?

By Zack Lau, W1VT
From *QST*, July 1991

Build a Portable Groundplane Antenna

Need a better antenna for your hand-held radio? Here's the answer.

The rubber ducky antennas common on hand-held VHF and UHF transceivers work fine in many situations. That's no surprise, considering that repeaters generally reside high and in the clear so you and your hand-held don't have to! Sometimes, though, you need a more efficient antenna that's just as portable as a hand-held. Here's one: A simple *groundplane* antenna you can build—for 146, 223 or 440 MHz—in no time flat. It features wire-end loops for safety (sharp, straight wires are hazardous) and convenience (its top loop lets you hang it off high objects for best performance).

What You Need to Build One

All you'll need are wire (single-conductor, no. 12 THHN), solder and a female coax jack for the connector series of your choice. Many hardware stores sell THHN wire—that is, thermal-insulation, solid-copper house wire—by the foot. Get 7 feet of wire for a 146-MHz antenna, 5 feet of wire for a 223-MHz antenna, or 3 feet of wire for a 440-MHz antenna.

The only tools you need are a 100-watt soldering iron or gun; a yardstick, long ruler or tape measure; a pair of wire cutters; a $1/2$-inch-diameter form for bending the wire loops (a section of hardwood dowel or metal tubing works fine), and a file (for smoothing rough cut-wire edges and filing the coax jack for soldering). You may also find a sharp knife useful for removing the THHN's insulation.

Building It

To build a 146-MHz antenna, cut three $24^5/8$-inch pieces from the wire you bought. To build a 223-MHz antenna, cut three $17^5/8$-inch pieces. To build a 440-MHz antenna, cut three $10^5/8$-inch pieces.

The photos show how to build the antenna, but they may not communicate why the cut lengths I prescribe are somewhat longer than the finished antenna's wires. Here's why: The extra wire allows you to bend and shape the loops by hand. The half-inch-diameter loop form helps you form the loops easily.

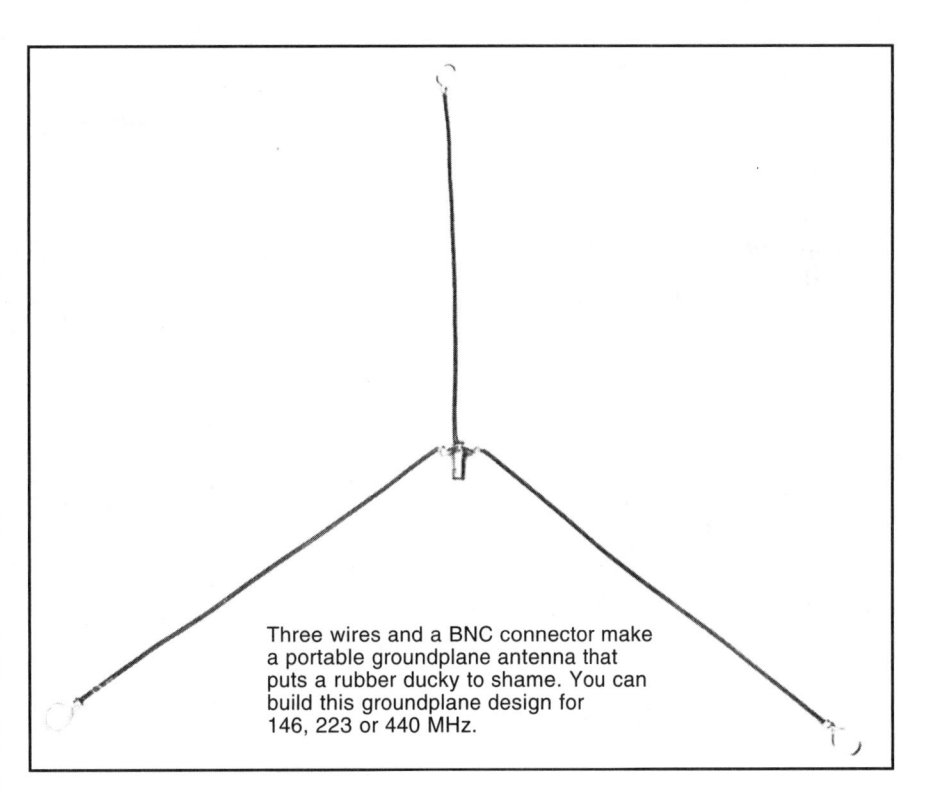

Three wires and a BNC connector make a portable groundplane antenna that puts a rubber ducky to shame. You can build this groundplane design for 146, 223 or 440 MHz.

Make the End Loops First

Form an end loop on each wire as shown in Fig 1. Strip exactly 4 inches of insulation from the wire. Using your $1/2$-inch-diameter form, bend the loop and close it—right up against the wire insulation—with a two-turn twist as shown in the bottommost example in Figure 1. Cut off the excess wire (about $1/2$ inch). Solder the two-turn twist. Do this for each of the antenna's three wires.

Attach the Vertical Wire to the Coax-Jack Center Pin

Strip exactly 3 inches of insulation from the unlooped end of one of your wires and follow the steps shown in Figure 2. Solder the wire to the connector center conductor. (Soldering the wire to a coaxial jack's center pin takes considerable heat. A 700- to 750-°F iron with a large tip, used in a draft-free room, works best. Don't try to do the job with an iron that draws less than 100 watts.) Cut off the extra wire (about $1/2$ inch).

Attaching the Lower Wires to the Connector Flange

Strip exactly 3 inches of insulation from the unlooped ends of the remaining two wires. Loop their stripped ends—right up to the insulation—through opposing mounting holes on the connector flange. Solder them to the connector. (You may need to file the connector flange to get it to take solder better.) Cut off the excess wire (about $2^1/4$ inches per wire). This completes construction.

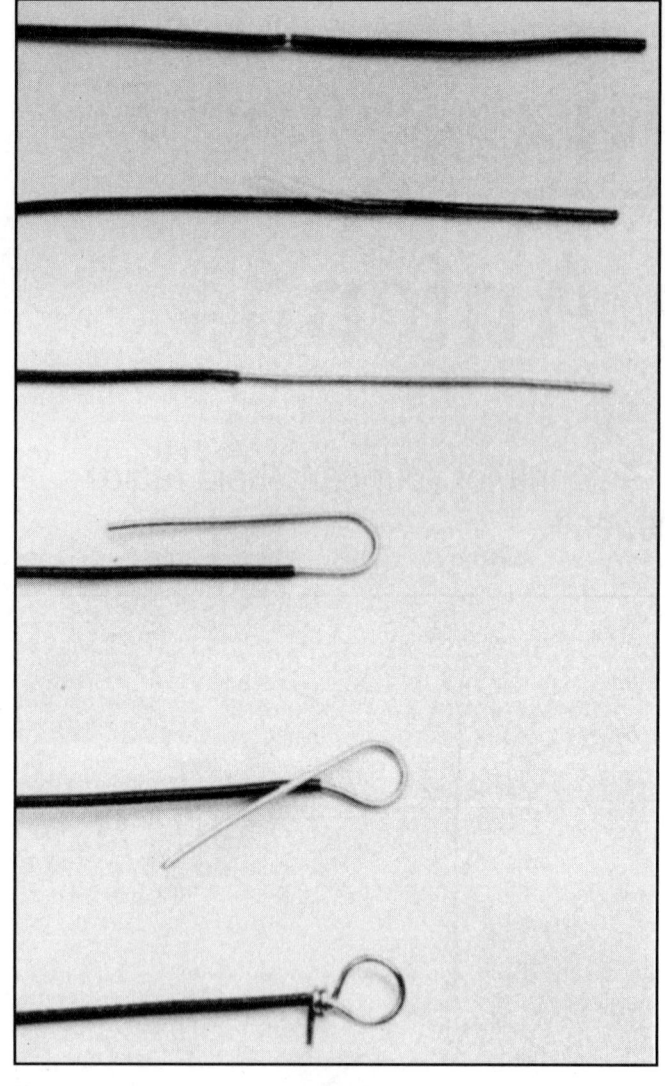

Figure 1—Making loops on the antenna wires requires that you remove exactly 4 inches of insulation from each. Stripping THHN insulation is easier if you remove its clear plastic jacket first.

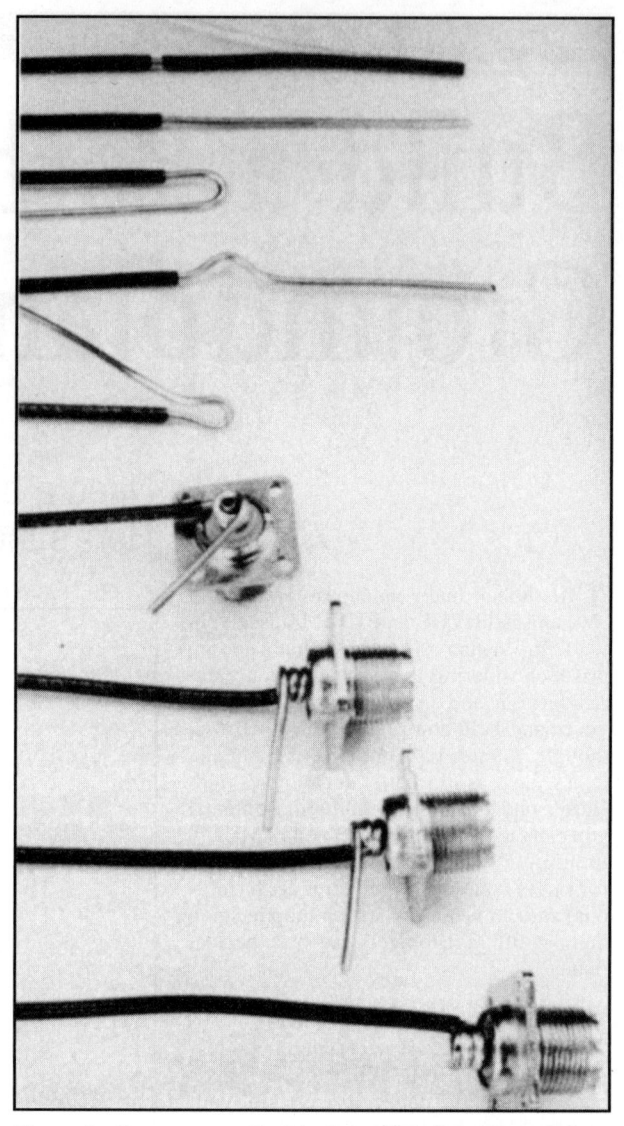

Figure 2—Remove exactly 3 inches of insulation to attach the vertical wire to the coax connector center pin. This photo shows an SO-239 (UHF-series) jack; the title photo shows a BNC jack. Use whatever your application requires.

Adjusting the Antenna for Best Performance

Bend the antenna's two lower wires to form 120° angles with the vertical wire. (No, you don't need a protractor: Just position the wires so they just about trisect a circle.) If you have no means of measuring SWR at your antenna's operating frequency, stop adjustment here and start enjoying your antenna! Every hand-held I know of should produce ample RF output into the impedance represented by the antenna and its feed line.

Adjusting the antenna for minimum SWR is worth doing if you have an SWR meter or reflected-power indicator that works at your frequency of interest. Connect the meter in line between your handheld and the antenna. Between short, identified test transmissions—on a simplex frequency—to check the SWR, adjust the angle between the lower wires and the vertical wire for minimum SWR (or reflected power). (You can also adjust the antenna by changing the length of its wires, but you shouldn't have to do this to obtain an acceptable SWR.) Before considering the job done, test the antenna in the clear to be sure your adjustments still play. (Nearby objects can detune an antenna.)

Plug and Play

As you use the groundplane, keep in mind that its coax connector's center pin wasn't made to bear weight and may break if stressed too much. Barring that, your groundplane should require no maintenance at all.

There you go: You may not have built a monument to radio science, but you've home-constructed a portable antenna that'll get much more mileage from your hand-held than its stock rubber ducky. Who said useful ham gear has to be hard or expensive to build?

> **What's a Groundplane?**
>
> This article emphasizes how to build and adjust a groundplane antenna for better communication at 146, 223 or 440 MHz. You can find out the technical details of how groundplane antennas work in Chapter 2 of *The ARRL Antenna Book*, available from your dealer and The ARRL Bookshelf.
> —WJ1Z

By E.J. Bauer, W9WQ

Constructing a Simple 5/8-Wavelength Vertical Antenna for 2 Meters

No loading coils—inexpensive—easy to build. Does that sound like the 5/8-λ antenna you've been wanting to build? You've got it now!

A diligent search of the amateur journals will reveal a plentiful supply of articles concerning the use of the deservedly popular 5/8-λ antenna on the VHF bands. Being a recent convert to 2-meter FM operation, I took a closer look at this type of antenna. In the process of constructing several versions, I formulated some new ideas which may be of interest to those who derive satisfaction from making their own antennas.

Electrical Theory

Refer to Fig. 1A. The feed-point impedance of a 5/8-λ vertical antenna exhibits a resistive component in the vicinity of 50 ohms. It requires a suitably chosen series inductor, however, to cancel the capacitive reactance which also exists at that point. Only then will a reasonable impedance match be presented to a 50-ohm coaxial-cable feed line. One constructor, K4LPQ, obtained the required inductance by means of a short-circuited stub of coaxial line of proper length.[1] (If the required reactance is known, the length of the stub can be calculated). He improved the mechanical construction of the antenna by placing the stub inside the radiating element. See Figure 1B. This approach requires an electrical connection to be made between the braid of the stub and the lower end of the radiator. Soldering such a connection would be difficult unless a material such as brass or copper is used for the radiator. If the center conductor of the stub is extended one-quarter wavelength beyond the inductive shorting point (as in Fig. 1C), a signal-frequency short will occur at that point. Furthermore, there is now no need for the stub to be made of coaxial cable. An insulated wire of suitable length (somewhat longer than an electrical quarter wavelength) is all that is needed to develop the required series inductance at the feed point of the antenna. It is only necessary to adjust the length of this stub until an acceptable VSWR is obtained. If desired, the radiator length can also be trimmed.

Construction

I selected a surplus whip antenna to be used as the radiating element for the 5/8-λ vertical.[2] Cutting the fully extended whip at a distance of four feet (1.22 m) from the tip leaves the larger (base) end with a 3/8-in. (10-mm) OD section. This size tubing fits closely into the hole at the barbed end of a PVC pipe adapter. See Figure 2. The hole in the threaded end of this same adapter mates snugly with the body of a

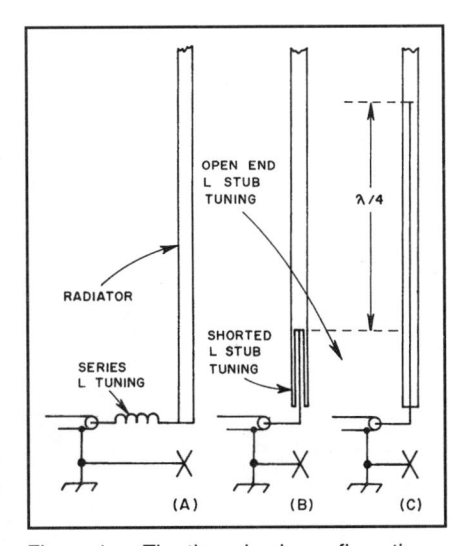

Figure 1 — The three basic configurations of the 5/8-λ antenna mentioned in the text. The stub tuning method shown at C may be easily arranged mechanically.

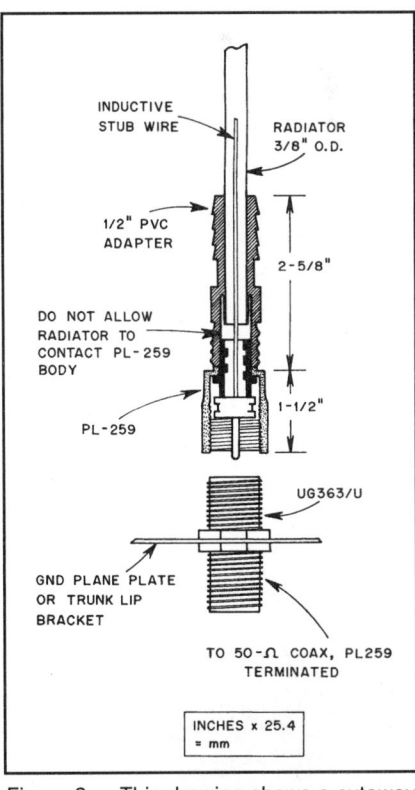

Figure 2 — This drawing shows a cutaway view of the antenna assembly. A PVC pipe adapter allows simple and inexpensive construction.

PL-259 coaxial connector. Cement the radiator to the adapter using a good adhesive. If epoxy is used, it would be advisable to roughen the inner surfaces of the plastic adapter to provide some "bite." It has been my experience that the bond between epoxy and PVC is marginal. Insert the radiator no more than 2 in. (51 mm) into the adapter and allow the adhesive to cure.

Solder approximately 28 in. (700 mm) of no. 18 solid, insulated wire to the pin of the PL-259. Larger wire may be used here, but the optimum length required for matching will be found to be somewhat longer. The inner diameter of this particular whip is $3/16$ in. (4.8 mm) at this point and will easily accommodate no. 12 insulated wire.

Figure 2 shows a UG-363/U connector being used as the junction for the radiator, ground plane and transmission line. This connector is expensive, so one might prefer to use the less expensive SO-239 connector.

Testing

Temporarily assemble the radiator/insulator assembly to the plug/wire portion, attach a ground plane and check the VSWR at two well-separated frequencies. The results will show whether the stub is too long or too short. It should be possible to get the VSWR below 1.5:1 across the repeater portion at the upper end of the 2-meter band. Shortening the stub to move the maximum VSWR point higher in frequency is easy. Should you overshoot, it is simple to start over again with a new piece of wire. Once you are satisfied with the results, the PL-259 can be cemented to the insulator. That's all there is to it. See you on 2 FM!

Notes
[1] Pentecost, "5/8-Wavelength Vertical Antenna for Mobile Work," *Ham Radio*, May 1976.
[2] Fair Radio Sales, P. O. Box 1105, 1016 E. Eureka St., Lima OH 45802, G01-51048 telescoping whip antenna.

By Bill English, N6TIW

From *QST*, April 1991

A Glass-Mounted 2-Meter Mobile Antenna

Want a no-holes, no-paint-scratching antenna? This easy-to-build glass-mounted mobile antenna is the answer!

To me, mag-mount antennas are a pain in the neck. Yet, I couldn't bear to drill a hole in my car to install a permanent mobile antenna. So, I found an alternative. In this article, I'll tell you how to build an attractive glass-mounted antenna that looks like those used for cellular-telephone service.

The antenna is easy to build. No special tools or skills are required, yet the antenna's appearance is compact and high tech. Instead of having an untidy mess on your car roof, people will think you are one of the nouveau riche. If you're not up to building one of the antennas, you can get one ready to install.[1]

The System

Figure 1 shows a schematic of a glass-mounted antenna system. At A is the electrical equivalent of a $1/4$-λ antenna. Why the need for L and C, and how does the signal get from the coax to the antenna without a connecting wire? The reason—and the way—is capacitance. The mounting plates on either side of the glass act as capacitor plates separated by a glass dielectric—your windshield or window. This capacitor exhibits a negative reactance. To cancel the negative reactance, inductance (L) is used. Because C and L have equal but opposite reactances at the frequency of interest, the feed line sees just the $1/4$-λ antenna. This antenna can be a simple $1/4$-λ whip, or it can be physically shortened with a loading coil to give it the cellular-phone look. More on this later.

Note that the feed-line shield is grounded near the capacitor. Because we're using a $1/4$-λ antenna, this is an important part of the system. Obviously glass makes a bad ground plane, but there is still plenty of metal around to do the job.

Antenna Design

The size of the antenna mount was

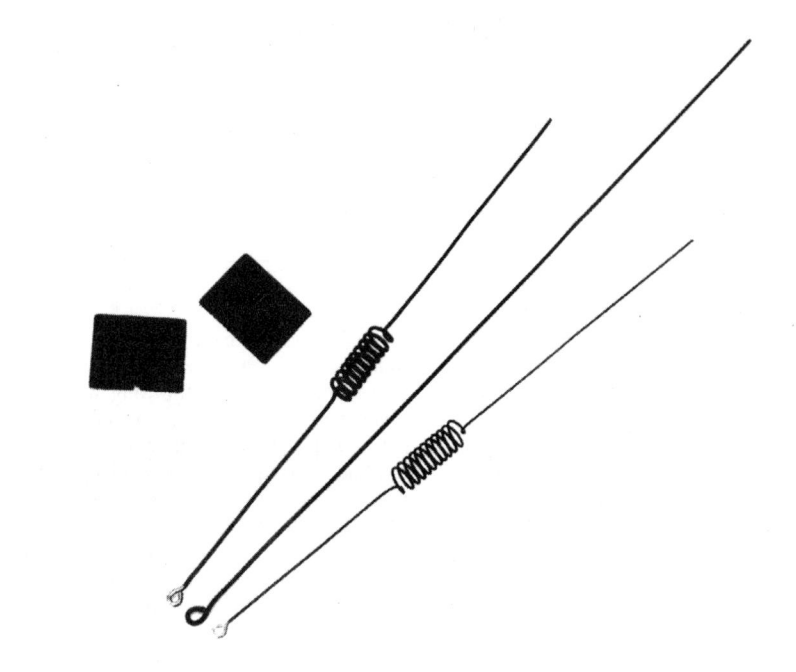

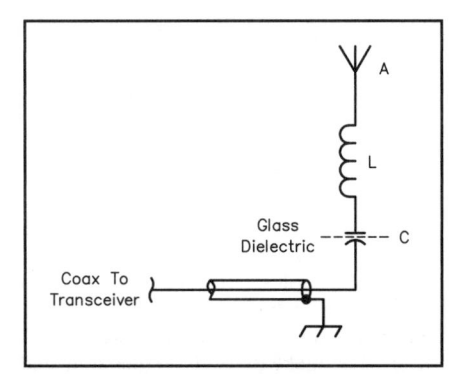

Figure 1—Schematic of a glass-mounted antenna system. The signal passes through the glass via the plates of the antenna mount, which form capacitor C. This capacitance is canceled by inductance L. Therefore, the feed line just sees the $1/4$-λ antenna A.

determined by structural and appearance considerations. I made the mount as big as my taste would allow. I calculated the mount's capacitance to be 10 to 20 pF. (I can't calculate it any closer because I'm uncertain about the dielectric constant and thickness of the glass and adhesive.) This capacitance range equates to a reactance of about 60 to 120 ohms at 2 meters. Therefore, the inductive reactance required to cancel the mount's capacitive reactance is 60 to 120 ohms.

How do you build the right amount of inductance into the antenna? Look at Figure 2. This graph is from *The ARRL Antenna Book* (Fig 36, p 2-36). When an antenna is shorter than 90 degrees, it looks capacitive to the feed line. When it's longer than 90 degrees, it looks inductive. Armed with this information, I went on two design paths:

144 MHz 2-29

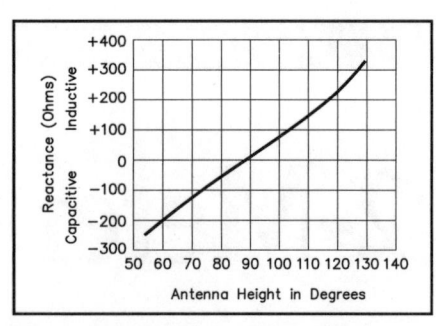

Figure 2—Approximate reactance of a vertical antenna over perfectly conducting ground. The wavelength/diameter ratio is about 2000. Actual reactance values vary considerably with wavelength/diameter ratio.

Figure 3—Antenna details. Finished dimensions are for my antennas using no. 10 copper wire. Start with extra length and trim for best SWR.

- Make a whip antenna longer than 90 degrees to provide the inductance needed to cancel the base mounting capacitance. From Figure 2, this is about 96-106 degrees, which calls for an antenna length of 20 to 22½ inches. Using no. 10 copper wire, my antenna ended up being 21½ inches long.
- Make a shortened antenna with a loading coil large enough to cancel the capacitive reactances of the base mounting capacitance and the shortened (less than 90 degrees) antenna length. I picked a 60-degree antenna length, which requires about 200 ohms of inductive reactance.[2] Adding an appropriate amount of inductance to cancel the base capacitance—and allowing some extra—I decided on a ½-inch diameter, 9-turn coil 2 inches long. This is all very approximate, but we don't have to worry too much about precision here. All sins will be forgiven when the antenna is trimmed.

Antenna Elements

I tried several materials for making antenna elements. Stainless steel is by far the most durable. It is hard to work in sizes over 1/16 inch, but 1/16-inch welding rod is workable with ordinary tools. Bare copper wire is by far the easiest to work, but it is too flexible unless first stresshardened by stretching.[3] (To ensure proper mounting-flange spacing, be sure to stretch the wire before making the antenna mount described in the next section.) Stress hardened copper wire stands up well to freeway driving. Feel free to experiment with other materials.

Figure 3 gives finished dimensions for antennas I've made. The length for best SWR may be slightly different when the antenna is mounted on your car, but that shouldn't matter because you'll be starting with a longer element length and trimming to size.

Making a long whip is easy. Cut a 24-inch length of straight wire. Bend one end into a small loop with needle-nose pliers. Be sure the loop will pass a no. 6 screw. Except for painting, you're done. Making the shortened antenna is only a little more complicated (see Figure 4). Sand the antenna wire before forming the coil. To make the coil, clamp one end of a 3-ft length of straight wire in a vise. Bend the free end into a small loop for the long whip. Then place the center of a ½-inch wooden dowel six inches from the center of the loop and roll the wire up on the dowel for 9 turns, moving toward the vise. Remove the wire from the vise and cut off the bent end. Hold one end of the coil with pliers and bend the free end so it cuts through the axis of the coil perpendicular to it. Then hold the bent portion with pliers near the coil edge and bend the wire so it is on the axis of the coil. Bend the free end on the other side of the coil along the coil's axis in the same way. Cut the end that was in the vise 7 inches or more from the coil. Space the turns so that the finished coil is about 2 inches long.

The Antenna Mount

The antenna mount (see Figs 5 and 6) is made from brass strips that are readily available at hobby stores. The details are shown in Figure 5. The inside plate is a 1½ × 2-inch piece of 0.016-inch brass with a 3/16-inch-wide tab cut from it. Brass of this thickness can be cut easily with metal shears. Just make two cuts ¼ inch long and bend the tab down. The tab is shown at the top of the plate in Figs 5 and 6, but it could be placed on the side if that better suits you. (The feed line will eventually be soldered to this tab.)

Now assemble the outside mount. Note that the right-angle flanges that hold the antenna are made from 1-inch-wide by 0.025-inch-thick brass strip. You'll need heavy-duty metal shears or a hacksaw to cut this brass. Start with 1 × 1-inch pieces and cut the corners off as shown in Figure 5. Then bend them by clamping them in a vise and hammering them over. Drill the hole in one flange before soldering to the mounting plate. To solder the flanges to the outside mounting plate, first apply solder to both parts. Then, press the flange onto the plate with a soldering iron to melt the solder. Before removing the iron, apply pressure with a screwdriver to keep the flange in place until the solder hardens. An iron with a reasonable amount of power is needed. (I used my 140-W gun.) After mounting the first flange, space the second flange from the first using the antenna wire. After both flanges are soldered in place, drill the second hole using the first hole as a guide. Be sure to file off all burrs, and for a neat appearance, round the corners and file off all solder blobs and flows. Set the mount aside for now.

Painting

Although it's not necessary to do so, I painted my mount and antenna. (*Don't paint the sides of the mounts that will be glued to the glass.*) Flat black paint provides a clean, finished look. Before painting, clean off all flux and sand surfaces

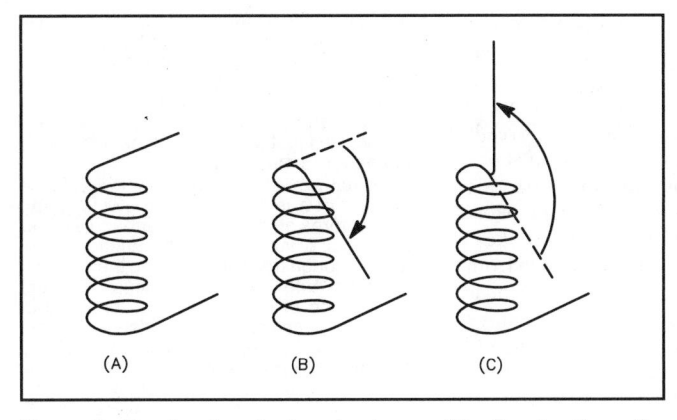

Figure 4—Bending the shortened antenna. After forming the coil (A), bond one free end so it crosses the coil axis perpendicular to it (B). Then bend up along the coil axis (C). Repeat this procedure on the other end so both ends are along the coil axis.

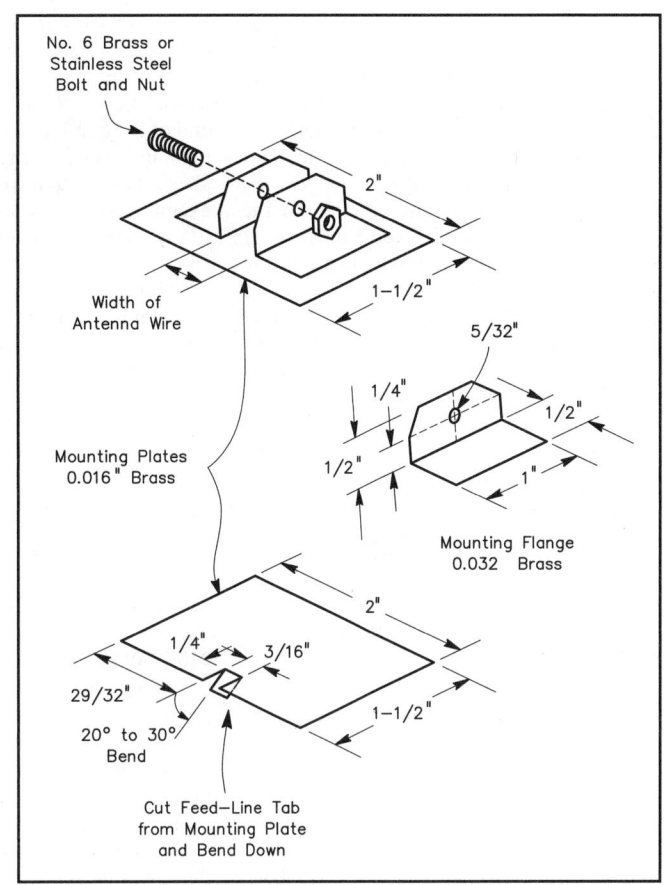

Figure 5—Antenna-mount details. Solder mounting flanges centered on the top mounting plate. The mounting flanges are cut from brass strips available at hobby stores.

with fine sandpaper until they are uniformly clean and bright. To paint, bolt the antenna in the outside mount and stand it upright. This makes it easier to paint and ensures that the electrical contact surfaces are not coated.

Brass doesn't take paint well. I primed my first prototype with auto primer. Although the paint looked fine, it chipped easily. An etching primer works much better. A primer designed for brass is best, but primers designed for aluminum will work too. Flecto Ferrothane Surfabond no. 52 is one of these. (Marine supply stores carry brass primer.) Once the metal is primed, apply a finish coat of a good quality flat black paint intended for metal. I used Rust-Oleum Bar-B-Q Black. It is a high-temperature paint that stands up to soldering.

Assembly

The most popular position for mounting on-the-glass antennas seems to be the top center of the rear window.[4] I have a station wagon, so this wasn't convenient for me. I located my antenna at the top center of the windshield. I like it in this position because the inside mount is neatly behind the rear view mirror and the coax route to the rig is short.

Before sticking the mount to the glass, solder the center conductor of the feed line to the inside plate's tab. Solder the coax braid to a ring lug, leaving enough braid exposed to reach a nearby screw in the headliner trim or mirror mount for grounding to the vehicle body. The grounding screw should be close to the mount, near the glass. If you don't have a screw conveniently close, put one there. The antenna won't work right without a properly grounded shield. If needed, drill a small hole in the headliner trim into the mounting metal, and insert a self-tapping or sheet-metal screw. After installing the mount, remove the screw, place the ring lug over it and replace the screw.

I tried several adhesives for the mount, but the easiest to use is double-sided foam tape. Radio Shack tape (RS 64-2361) ap-

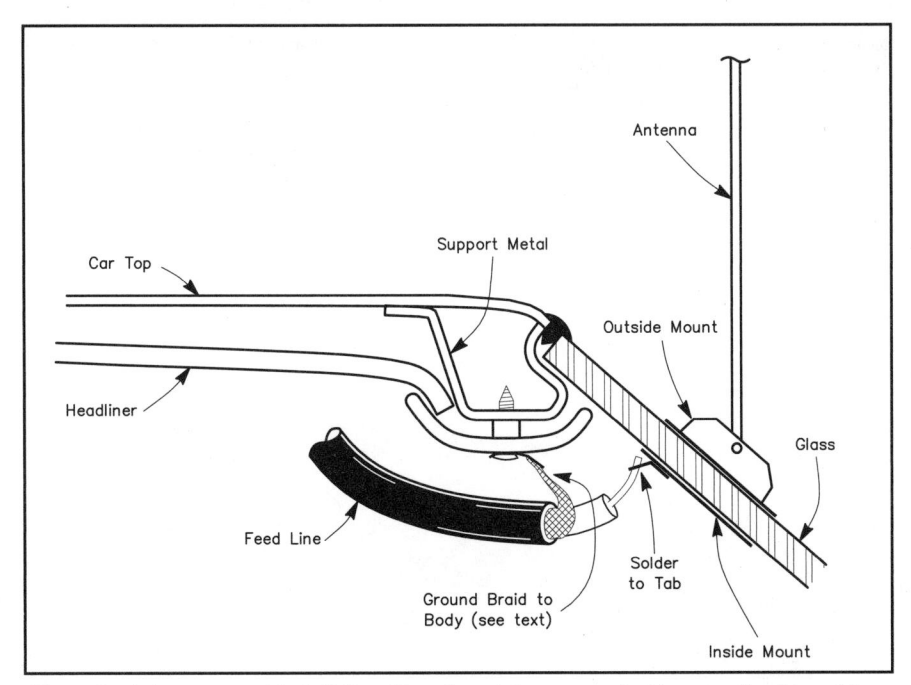

Figure 6—Antenna-mount installation. Mounting plates are applied to opposite sides of the glass. The antenna is bolted to the outside mount. The center conductor of the feed line is soldered to the inside mount. Ground the coax braid to the body using a nearby screw.

144 MHz 2-31

pears to be Scotch no. Y-4950, which is described as having "high peel strength and excellent weatherability." The tape comes in 1-inch-wide strips so use two strips side by side on each plate. Clean the glass and the mounting plates with rubbing alcohol before application. Apply tape to the mounting plate first, then stick it to the glass. Once it's applied, don't move it. If you make a mistake, remove the mounting plate, clean it up and put on a new adhesive strip. Although this tape should weather well, it's a good idea to seal around the edges of the outside plate with a thin bead of a silicone sealer (such as Radio Shack 64-2314).

Use RG-58 cable for the feed line if you mount the antenna on the windshield. This small-diameter coax is easily routed to the transceiver. RG-58 has rather high loss at 2 meters. For the lengths used here, however, losses are below 1 dB, so they won't be noticeable. For line lengths over 10 or 15 feet, use RG-8X (Radio Shack RG-8M) coax. It has a slightly larger diameter than RG-58, but losses are lower.

Routing the feed line to your rig requires some ingenuity if you want the feed-line run to be invisible, or almost so. If you install the antenna on the vehicle's rear window, you may want to route the coax under the headliner. Look for screws or other headliner retainers around the trim edges. You should be able to drop some of the headliner and snake the coax through. I tucked the feed line into a gap between the headliner and the windshield trim along the top of the windshield, then used small dabs of cement to run it in an indentation along the door post to my transceiver. This worked well because my car's black trim matched the coax outer covering. Terminate the transceiver end of the feed line with a connector to match the one on your rig. If you're going to use a hand-held transceiver, cut the coax a bit long.

Pruning the Antenna

Once the antenna is installed, trim it for minimum SWR. Insert an SWR meter between the rig and the feed line. With your rig set for low power, check the SWR near the bottom and top ends of the 2-meter band. The SWR should be lower at the low end of the band. If not, and both SWR readings are over 2:1, the antenna is too short. There isn't much you can do to fix this except make a new antenna element. Most likely, however, everything will be okay. Using wire cutters, trim about 1/8 inch from the antenna tip, rechecking the SWR near the center of the band. It's best to remove only 1/8 inch at a time to avoid cutting off too much. Keep trimming the antenna length and checking the SWR until you get it near 1:1. Check the SWR near the band edges and at your usual frequencies of operation. Trim carefully if you want better SWR at the higher end of the band. I trimmed my antenna for lowest SWR at 146 MHz. At the band edges, the SWR of my antenna is 1.4:1, giving good performance over the whole band.

Summary

There you have it! With a glass-mounted antenna, there is no need to drill holes through your roof or put up with a magmount mess. A glass-mounted antenna gives you convenience and a high-tech look.

My operating experience confirms that these antennas perform about the same as a regular $1/4$-λ antenna. The long whip should have slightly more gain than the shortened one, but both should be within 1 dB of a roof-mounted $1/4$-λ antenna. This is plenty good unless you are in a fringe area. In fact, higher-gain antennas won't necessarily help you if you are in a "hole" in hilly terrain. This situation is common where I live in the San Francisco Bay area. I have been very pleased with the shortened antenna's performance and appearance.

Notes

[1] Contact William J. English, 81 Meadow View Rd, Orinda, CA, 94563. (The ARRL and QST in no way warrant this offer.)

[2] Actually, this coil-sizing method applies only to a base-loaded antenna with a coil small enough that its radiation can be neglected. Our antenna is center loaded (this increases the inductance required) with significant radiation (reduces inductance required). Rather than sorting these effects out mathematically, I just made the element long and trimmed for best SWR.

[3] Here's how to stretch no. 10 copper wire. Place one end of a length of wire in a vise. Wrap several turns of the other end around a crowbar about 6 inches from the fulcrum end. Leave enough free wire between the vise and bar for the antenna you are making. Place the fulcrum end of the bar against the edge of your work bench opposite the vise, or against another solid object. Pull until you feel the wire give several inches. Not only will this stiffen the wire, but it also straightens out all the kinks. Cut the wire at the crowbar end. Leave the straight wire clamped to the vise to form the coil if you are making the shortened antenna.

[4] A letter from Joseph Butcher, KE9FZ, prompted your editor to seek some information on the RF properties of tinted window glass. Mounting on glass antennas to factory-tinted (deep-tint) windows, he says, results in problems with high SWR.

Rex Greenshade of the Ford Motor Company told me that he believes OEM-specified tinted glass does not cause the problem: Aftermarket heavily tinted glass is the gremlin. Rex said that the maximum OEM glass tinting must conform to certain specifications, one reason being that police officers must be able to view the occupants of the vehicle when approaching on foot.

Ken Brown of the Ford Motor Company told me that no Ford OEM glass tints interfere with RF. However, Instaclear—not a tint—does interfere with the transmission of AF. Radar cannot penetrate it and it does interfere with the installation of glass-mount antennas. Instaclear is a Ford option that permits rapid clearing of ice and snow from a car's windshield. Instaclear glass has a conductive powder added to the layers of windshield glass. Although Instaclear is not designed as a tinting element, under certain lighting conditions and viewing angles it can appear as a pink or bronze tint that is highly reflective and virtually opaque. Ken said that Instaclear does not violate any existing state laws governing tinted window glass.

Scott Staedtler, N8ILG, of Antenna Specialists Co. told me that in his experience, the degree of tinting does indeed have an effect on glass-mounted antennas. He's aware of Instaclear, but claims that the OEM privacy glass tint (as used on Ford vans and Broncos, for instance) as well as aftermarket tints can cause problems. Rear-window defoggers/deicers do not interfere with on-glass antennas so long as you avoid mounting the antenna on the wire trace(s).—Ed.

By Dennis Blanchard, K1YPP

From *QST*, July 1995

Build a Weatherproof PVC J-Pole Antenna

After you've built this antenna, you may not need anything else.

The twinlead J-pole antenna has been around for quite some time. It was brought into the limelight by an excellent article written by John S. Belrose, VE2CV, in the April 1982 *QST*. While John provided an excellent theoretical discussion of the J-pole, his article did not offer great detail on precisely how to *build* this wonderful VHF/UHF antenna.

J-poles are easy to build—which is why you see so many versions in use. (And so many articles in print!) Even so, several misconceptions exist concerning the J-pole. One common mistake is to assume that all you have to do is attach a piece of coaxial cable to a length of twinlead, short the bottom section and cut a notch. Not quite!

Another misconception is that once the antenna is built and tuned, you can stuff it inside a PVC tube and expect it to work flawlessly. Unfortunately for many amateurs, the PVC treatment often results in a failed antenna—unless you do it *right*.

Understanding J-Pole Construction

The J-pole antenna comprises two parts (see Figure 2): a ¼-wavelength matching section, which is the *entire portion below the notch*; and the radiating section, which is the ½-wavelength section *above* the notch. The portion of the antenna below the notch is most affected by the type of insulation that surrounds it. It also has the most influence on the resonance of the antenna. The radiating section is not as greatly affected by the insulation or the type of wire used. (We'll discuss this effect in a moment.)

When installed inside a PVC tube, the J-pole is a rugged and weather resistant antenna. If you place a J-pole inside PVC, however, you must *center the antenna within the tube*. One way to do this is to place the antenna inside a piece of foam insulation, preferably the type used to insulate hot-water pipes, before you slide it into the tube. If you choose a 1.5-inch PVC tube, this insulation is often a perfect fit (see Figure 1).

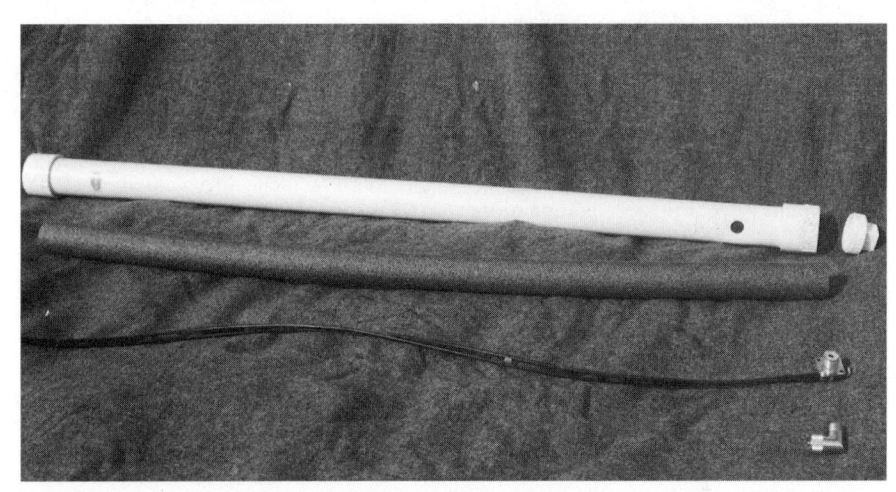

The disassembled J-pole antenna. The twinlead antenna core is shown at the bottom with the M359 right angle connector removed. It's placed within a foam insert (middle) which keeps the antenna centered within the PVC tube (top). This is the construction technique used by the JADE Products "JADE-POLE" antenna.

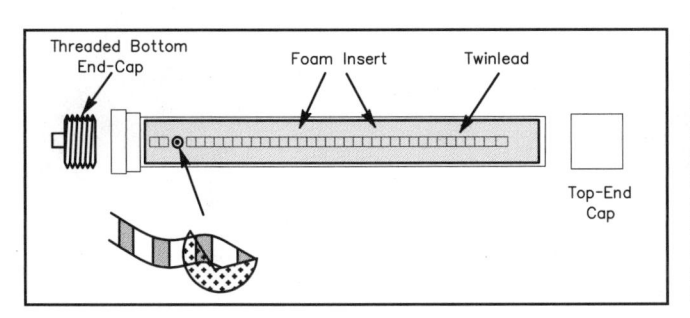

Figure 1—In this cut-away view you can see that a foam insert keeps the antenna centered within the PVC tube. End caps keep out moisture and an M-359 right-angle connector makes it easy to attach the coax.

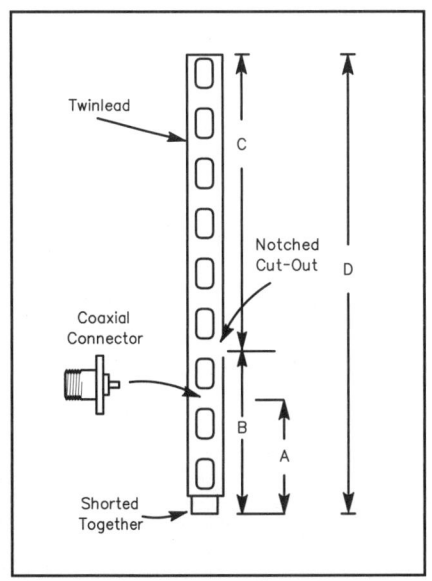

Figure 2—The critical lengths for the J-pole (see Table 1). Note the notch that's cut into one of the twinlead wires. The wires at the bottom are shorted together.

144 MHz 2-33

Table 1
Section Lengths (See Figure 3)

Frequency (MHz)	D, Total Length (in.)	A (in.)	B (in.)	C (in.)
50.00	160.4	3.2	48.2	112.2
51.00	157.2	3.1	47.2	110.0
52.00	154.2	3.1	46.3	107.9
53.00	151.3	3.0	45.5	105.9
54.00	148.5	3.0	44.6	103.9
146.00	54.9	1.1	16.5	38.4
222.00	36.1	0.7	10.9	25.3

Tuning Your 6-Meter J-Pole

You can use a capacitive coupling strap to easily tune your 6-meter J-pole for a different portion of the band. No cutting or lengthening necessary!

You can make the strap from a 1-inch wide strip of aluminum foil. Wrap the foil around the lower section of the J-pole and hold it in place with electrical tape. The strap doesn't connect to the antenna. It merely increases the capacitance at that point where it's positioned. By moving the strap up and down along the lower section, you'll change the resonant frequency of the antenna. This technique works best on a J-pole designed for 50 MHz (See Table 1).

Building a J-Pole Antenna

STEP ONE: The Decision Phase

Choose a frequency for your J-pole. In the case of 144 or 220 MHz bands, the antenna bandwidth is many megahertz, so this isn't a critical decision. Simply use the middle of the band, 146 MHz and 222 MHz, respectively. However, on 50 MHz the antenna will not cover the entire band without readjustment. On 50 MHz the bandwidth will be approximately 2 MHz. This means you'll need to select a frequency that corresponds to your favorite portion of the band.

Table 1 gives you the cutting lengths for the antenna sections. But before you can start cutting, you need to consider the *velocity factor* of the twinlead you're using. Despite what you may have heard, RF energy does not flow through a cable at the speed of light in a vacuum. The wire and even the insulation act to slow the speed of the wave. So, the time required for the signal to travel through a length of cable is *longer* than the time required to travel the same distance in free space. This means that the full wavelength of the signal exists in a physically *shorter* length of cable. If you cut the cable for the wavelength of the signal in free space, you'll be off the mark!

Cable manufacturers test for the velocity factor and specify it as a decimal percentage of the speed of light. The lengths shown in Table 1 are based on windowed 300-Ω twinlead with a velocity factor of 0.85. If other twinlead is used, you may need to increase or decrease the lengths proportionally. For example, if a section length is 16$^{1}/_{2}$ inches long and you're using TV twinlead with a typical velocity factor of 0.83, reduce the length by 2%, to 16$^{3}/_{16}$ inches. (A velocity factor of 0.83 is roughly 98% of 0.85. Putting it another way, it's 2% *less* than 0.85.)

Next, decide how the antenna will be used: indoors or outdoors, fixed station or portable. If the antenna is to be used indoors, weather sealing will not be needed. If you're going to use it outdoors, apply a sealant to cover the exposed metal (the coaxial cable connection and the copper wire in the twinlead).

To limit possible RF absorption, use schedule-40 PVC. Make sure it is ultraviolet resistant as well.

Applying a sealant directly to the twinlead will change the resonant frequency of the antenna. At first this may seem a bit odd. But, believe it or not, the sealant *does* affect the velocity factor of the twinlead. If the velocity factor changes, the resonant frequency of the antenna changes. Usually it will be lower than calculated. For example, an antenna cut for 146 MHz may resonate at 142 MHz after the exposed conductors are coated with sealant—a 4% change!

STEP TWO: Cutting the Wire

Select a good grade of 300-Ω twinlead, one that is tough and will withstand abuse. Avoid TV-grade twinleads that tend to crack easily. Windowed 300-Ω twinlead is available from several *QST* advertisers.

Measure a length of twinlead that is approximately 10% longer than the amount needed. Measure it so that the notch will be cut where there is insulation all the way across between the two conductors, not at a "window."

Cut the notch. Cut only one wire; the other will run the full length of the antenna. The notch can be a small V or square. Make it at least a $^{1}/_{4}$ inch long. Measure from the notch to the bottom of the antenna cut off the excess wire. Strip about $^{1}/_{4}$ inch of in-

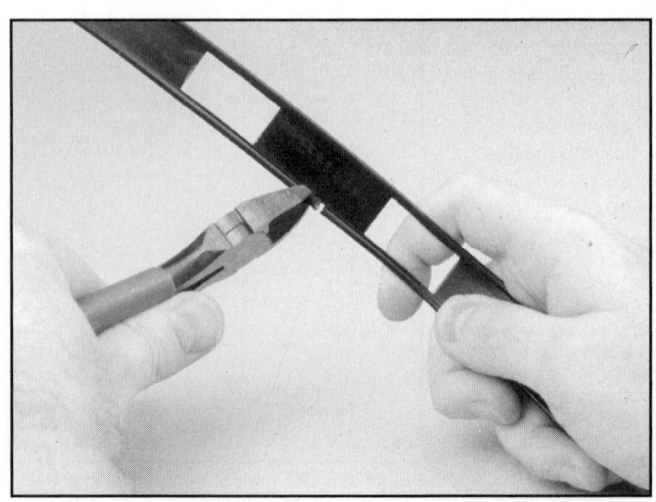

Cut the notch in only *one* of the twinlead wires. The twinlead shown in these photographs is 450 Ω. However, the same techniques apply to 300-Ω twinlead.

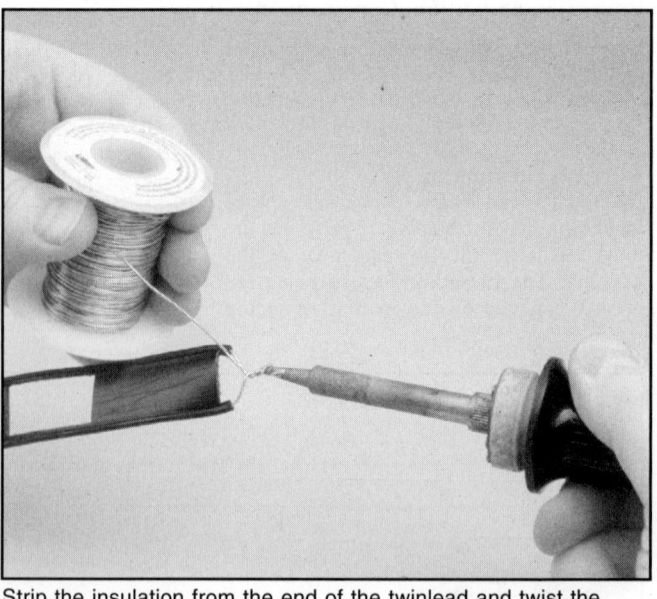

Strip the insulation from the end of the twinlead and twist the conductors together. A little solder ensures a good electrical connection.

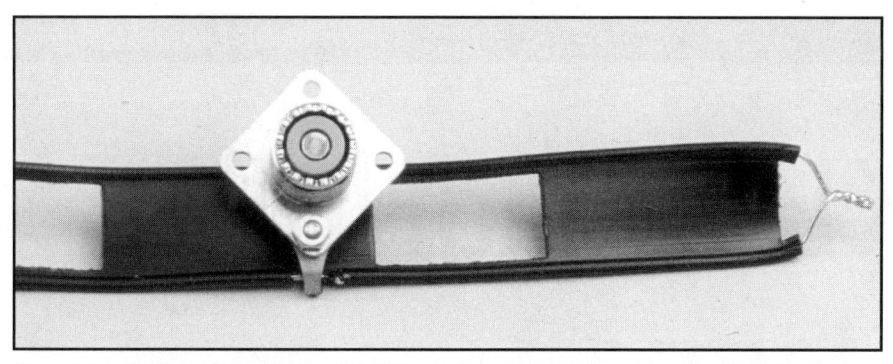

The center pin (not shown) of the SO-239 coaxial connector is soldered to the wire that runs the full length of the antenna. You can use a file to notch the pin. This will make it easier to solder. Then, attach a solder lug to the exterior of the SO-239 using the appropriate screw and nut. Strip away enough insulation to expose the wire on the opposite side of the twinlead and solder the lug in place.

sulation off each of the wires at the bottom. Take a small piece of bare wire and wrap several turns between the two exposed wires at the bottom. Now measure from the bottom to the top of the antenna and cut off the excess. Using a razor knife or other sharp knife, remove the insulation where the coax will be connected.

STEP THREE: Connect the Coax

You have two choices: You can solder the coaxial cable directly to the twinlead, or install a UHF or BNC coaxial connector. A connector is highly recommended because it allows quick connections and disconnections. It also provides some strain relief, so the connection is less likely to break.

If you decide to use a connector, first file a slot in the center conductor of the connector and set the antenna wire into the slot. For the other connection, mount a solder lug on one of the holes on the connector. Wrap the lug around the wire, or slot the lug and slip the wire into it. Solder both conductors. One word of caution: Make sure the center conductor is connected to the wire that runs the full length of the antenna and that the braid side of the coax is connected to the notched side.

After you install your J-pole in a PVC tube, an M-359 right-angle coax connector comes in handy. It makes it much easier to bring the coax connection outside the tube. You can create a flat spot on the tube with a heat gun. Heat the PVC carefully until it softens, then press down with a narrow piece of wood. By creating this flat surface and using a small rubber gasket, you'll have a waterproof seal for the coax connector.

STEP FOUR: Test the Antenna

You can tune the antenna with an SWR analyzer, if you have one, or just an accurate SWR meter. The resonant frequency of your J-pole is where you'll find the lowest SWR. The 144 and 220-MHz versions have a bandwidth almost twice as wide as the bands themselves, so tuning should not be necessary. The 50-MHz version may require minor tuning to make it resonant at the correct frequency (see the sidebar, "Tuning Your 6-Meter J-Pole"). Place the antenna in the foam core and PVC *before* you check for resonance.

If you find that you need to tweak your J-pole, make the matching section at the bottom slightly longer. Usually this will not be necessary.

STEP FIVE: Installation

You can install your PVC J-pole on a mast, or against a flat nonconductive wall. Plastic clamps for 1.5 and 1.0-inch PVC are available from JADE Products, PO Box 368, East Hampstead, NH 03826. Consider drilling a tiny hole in the bottom of the tube to allow any water to escape.

Conclusion

In his original work, VE2CV recommended placing a choke near the coaxial connection. To fashion a simple choke, take a cylindrical ferrite (Amidon 2X-43-251) and attach it to the coax at the feed point.

The J-pole antenna does not need radials, so it has a very narrow profile and low wind resistance. This is particularly important if you live in an area where icing is a problem. If the PVC enclosure has a threaded bottom, the antenna can be attached to a short piece of mating PVC and mounted above surrounding surfaces.

By Eugene F. Ruperto, W3KH

From *QST*, August 1996

The W3KH Quadrifilar Helix Antenna

If your existing VHF omnidirectional antenna coverage is "just okay," this twisted 'tenna is probably just what you need!

I still remember that hollow, ghostly signal emanating from my receiver in 1957. The signal was noisy and it faded, but that was to be expected—it was coming from outer space. I couldn't help but marvel that mankind had placed this signal sender in space! They called it *Sputnik,* and it served to usher in the space race.

Little did I realize then that four decades later we would have satellites in orbit around Earth and other heavenly bodies performing all sorts of tasks. Now we tend to take satellites for granted. According to the latest information on the Amateur Radio birds, I count about 15 low-Earth-orbit (LEO) satellites for digital, experimental and communications work, and two in *Molniya*-type highly elliptical orbits (AO-10 and AO-13), with the probability of a third to be launched in early 1997.

The world has access to several VHF weather satellites in low Earth orbit. Unlike geostationary Earth-orbiting satellites (GOES), the ever-changing position of the LEOs presents a problem for the Earth station equipped with a fixed receiving antenna: signal fading caused by the orientation of the propagated wavefront. This antenna provides a solution to the problem. Although this antenna is designed primarily for use with the weather sats, it can also be used with any of the polar-orbiting satellites.

These days, technical advances and miniature solid-state devices make it relatively easy for an experimenter to acquire a weather-satellite receiver and a computer interface at an affordable price. So it was only a matter of time before I replaced my outdated weather-sat station with state-of-the-art equipment.

Yesterday

In the early '70s, I built a drum recorder that used a box with a light-tight lid. It was a clumsy affair. The box and photo equipment took up most of the 6×8-foot room in which it was housed. Next to the recorder, a 3×4-foot table supported a tube-type receiver, frequency converters, a reel-to-reel tape recorder (our data-storage medium), a 50-pound monitor oscilloscope, az/el rotator controls for the helical antennas and a multitude of other devices including the drum-driver amplifiers and homemade demodulator. This station provided coverage of the polar-orbiting and geostationary satellites and furnished me with "tons" of data. Over time, my weather-satellite station evolved into a replica of mission control for the manned-spaceflight program! I had so much gear, it had to be housed in a shed separate from the house.

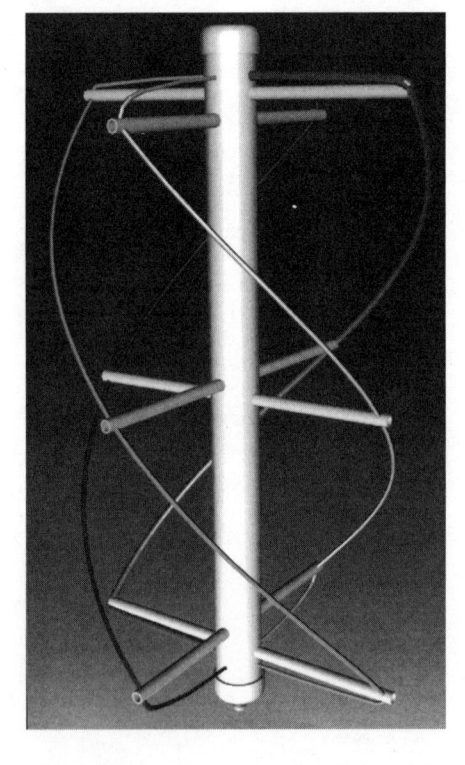

Today

Now, my entire weather-satellite station sits unobtrusively in one corner of the shack, occupying an area of less than one square foot—about the same size as my outboard DSP filter. My PC—now the display for weather-sat photos—is used for many applications, so an A/B switch allows me to toggle the PC between the printer and the weather-satellite interface.

What I needed next was a simple antenna system for unattended operation—something without rotators—something that would provide fairly good coverage, from about 20° above the horizon on an overhead pass. It was a simple request, but apparently one without a simple solution.

Background

Initially I used a VHF discone antenna with mixed results. The discone had a good low-elevation capture angle, but exhibited severe pattern nulls a few minutes after acquisition of signal and again when the satellite was nearly overhead. The fades and nulls repeated later as it approached the other horizon. About this time, Dave Bodnar, N3ENM (who got me reinterested in the antenna project), built a turnstile-reflector (T-R) array. The antenna worked fairly well but exhibited signal dropout caused by several nulls in the pattern. Dave built two more T-Rs, relocating them for comparison purposes. Unfortunately, the antennas retained their characteristic fades and nulls. Another experimenter and I built T-Rs and we experienced the same results. I suggested that we move on to the Lindenblad antenna. The Lindenblad proved to be a much better antenna for our needs than either the T-R or the discone,

Figure 1—The humble beginnings of a terrific antenna.

Figure 2—The quadrifilar helix antenna with two of the four legs (filars) of one loop attached.

Figure 3—This view shows the QHA with all four legs in place. The ends of the PVC cross arms that hold the coaxial leg are notched; the wire elements pass through holes drilled in the ends of their supporting cross arms.

but still exhibited nulls and fades. Over a period of several months, I evaluated the antennas and found that by switching from one antenna to another on the downside of a fade, I could obtain a fade-free picture, but lost some data during the switching interval. Such an arrangement isn't conducive to unattended operation, so my quest for a fade-free antenna continued.

The Quadrifilar Helix Antenna

Several magazines have published articles on the construction of the quadrifilar helix antenna (QHA) originally developed by Dr. Kilgus,[1] but the articles themselves were generally reader unfriendly—some more than others. One exception is *Reflections* by Walt Maxwell, W2DU.[2] Walt had considerable experience evaluating and testing this antenna while employed as an engineer for RCA.

Part of the problem of replicating the antenna lies in its geometry. The QHA is difficult to describe and photograph. Some of the artist's renditions left me with more questions than answers, and some connections between elements as shown conflicted with previously published data. However, those who have successfully constructed the antenna say it is *the* single-antenna answer to satellite reception for the low-Earth-orbiting satellites. I agree.

Design Considerations

I had misgivings about the QHA construction because the experts implied that sophisticated equipment is necessary to adjust and test the antenna. I don't disagree with that assumption, but I *do know* that it's possible to construct a successfully performing QHA by following a cookbook approach using scaled figures from a successful QHA. These data—used as the design basis for our antennas—were published in an article describing the design of a pair of circularly polarized S-band communication-satellite antennas for the Air Force[3] and designed to be spacecraft mounted. Using this antenna as a model, we've constructed more than six QHAs, mostly for the weather-satellite frequencies and some for the polar-orbiting 2-meter and 70-centimeter satellites with excellent results—*without the need for adjustments and tuning*. Precision construction is not my forte, but by following some prescribed universal calculations, a reproducible and satisfactory antenna can be built using simple tools. The proof is in the results.

The ultrahigh frequencies require a high degree of constructional precision because of the antenna's small size. For instance, the antenna used for the Air Force at 2.2 GHz has a diameter of 0.92 inch and a length of 1.39 inches! Nested inside this helix is a smaller helix, 0.837 inch in diameter and 1.27 inches in length. In my opinion, construction of an antenna *that* size requires the skill of a watchmaker! On the other hand, a QHA for 137.5 MHz is 22.4 inches long and almost 15 inches in diameter. The smaller, nested helix measures 20.5 by 13.5 inches; for 2 meters, the antenna is not much smaller. Antennas of this size are not difficult to duplicate even for those of us who are "constructionally challenged" (using pre-cut pieces, I can build a QHA in *less than an hour!*).

Electrical Characteristics

A half-turn half-wavelength QHA has a theoretical gain of 5 dBi and a 3-dB beamwidth of about 115°, with a characteristic impedance of 40 Ω. The antenna consists basically of a four-element, half-turn helical antenna, with each pair of elements described as a *bifilar,* both of which are fed in phase quadrature. Several feed methods can be employed, all of which appeared to be too complicated for us with the exception of the infinite-balun design, which uses a length of coax as one of the four elements. To produce the necessary 90° phase difference between the bifilar elements, either of two methods can be used. One is to use the same size bifilars, which essentially consist of two twisted loops with their vertical axes centered and aligned, and the loops rotated so that they're 90° to each other (like an egg-beater), and using a quadrature hybrid feed. Such an antenna requires *two* feed lines, one for each of the filar pairs. The second and more practical method, in my estimation, is the self-phasing system, which uses *different-size loops:* a larger loop designed to resonate *below* the design frequency (providing an inductive reactance component) and a smaller loop to resonate *higher* than the design frequency (introducing a capacitive-reactance component), causing the current to lead in the smaller loop and lag in the larger loop. The element lengths are 0.560 λ for the larger loop, and 0.508 λ for the smaller loop. According to the range tests performed by W2DU, to achieve *optimum* circular polarization, the wire used in the construction of the bifilar elements should be 0.0088 λ in diameter. Walt indicates that in the quadrifilar mode, the

Figure 4—Another view of the QHA.

Table 1
Quadrifilar Helix Antenna Dimensions

Freq (MHz)	Wavelength λ (inches)	Leg Size (0.508 λ)	Diameter (0.156 λ)	Small Loop Length (0.238 λ)	Leg Size (0.560 λ)	Big Loop Diameter (0.173 λ)	Length (0.26 λ)
137.5	85.9	43.64	13.4	20.44	48.10	14.86	22.33
146	80.9	41.09	12.6	19.25	45.30	14.0	21.03
436	27.09	13.76	4.22	6.44	15.17	4.68	7.04

Figure 5—An end-on view of the top of the QHA prior to soldering the loops and installing the PVC cap.

fields from the individual bifilar helices combine in optimum phase to obtain unidirectional end-fire gain. The currents in the two bifilars must be in quadrature phase. This 90° relationship is obtained by making their respective terminal impedances $R + jX$ and $R - jX$ where $X = R$, so that the currents in the respective helices are −45° and +45°. The critical parameter in this relationship is the terminal reactance, X, where the distributed inductance of the helical element is the primary determining factor. This assures the ±45° current relationship necessary to obtain true circular polarization in the combined fields and to obtain maximum forward radiation and minimum back lobe. Failure to achieve the optimum element diameter of 0.0088 λ results in a form of elliptical, rather than true circular polarization, and the performance may be *a few tenths of a decibel* below optimum, according to Walt's calculations. For my antenna, using #10 wire translates roughly to an element diameter of 0.0012 λ at 137.5 MHz—not ideal, but good enough.

To get a grasp of the QHA's topography, visualize the antenna as consisting of two concentric cylinders over which the helices are wound (see Figures 1 through 5). In two-dimensional space, the cylinders can be represented by two nested rectangles depicting the height and width of the cylinders. The width of the larger cylinder (or rectangle) can be represented by 0.173 λ, and the width of the smaller cylinder represented by 0.156 λ. The length of the larger cylinder or rectangle can be represented by 0.260 λ, and the length of the smaller rectangle or cylinder can be represented by 0.238 λ. Using these figures, you should be able to scale the QHA to virtually any frequency. Table 1 shows some representative antenna sizes for various frequencies, along with the universal parameters needed to arrive at these figures.

Physical Construction

After several false starts using plywood circles and plastic-bucket forms to hold the helices, I opted for a simple PVC solution that not only is the simplest from a constructional standpoint, but also the best for wind loading. I use a 25-inch-long piece of schedule 40, 2-inch-diameter PVC pipe for the vertical member. The cross arms that support the helices are six pieces of ½-inch-diameter PVC tubing: three the width of the large rectangle or cylinder, and three the width of the smaller cylinder. Two cross arms are needed for the top and bottom of each cylinder. The cross arms are oriented perpendicularly to the vertical member and parallel to each other. A third cross arm is placed midway between the two at a 90° angle. This process is repeated for the smaller cylindrical dimensions using the three smaller cross arms with the top and bottom pieces oriented 90° to the large pieces. Using ⅝-inch-diameter holes in the 2-inch pipe ensures a reasonably snug fit for the ½-inch-diameter cross pieces. Each cross arm is drilled (or notched) at its ends to accept the lengths of wire and coax used for the elements. Then the cross arms are centered and cemented in place with PVC cement. For the weather-satellite antennas, I use #10 copperclad antenna wire for three of the helices and a length of RG-8 for the balun, which is also the fourth helix. (I do not consider the velocity factor of the coax leg for length calculation.) For the UHF antennas, I use #10 soft-drawn copper wire and RG-58 coax. Copperclad wire is difficult to work with, but holds its shape well. Smaller antennas can be built without the cross arms because the wire is sufficiently self-supporting.

To minimize confusion regarding the connections and to indicate the individual legs of the helices, I label each loop or cylinder as B (for big) and S (for small); T and B indicate top and bottom. See Figures 6 and 7. I split each loop using leg designators as B1T and B1B, B2T and B2B, S1T and S1B and S2T and S2B, with B2 being the length of coax and the other three legs as wires. For right-hand circular polarization (RHCP) I wind the helices *counterclockwise* as viewed from the top. This is contrary to conventional axial-mode helix construction. (For LHCP, the turns rotate *clockwise* as viewed from the top.) See Figure 7 for the proper connections for the top view. When the antenna is completed, the view shows that there are two connections made to the center conductor of the coax (B2) top. These are B1T and S1T, for a total of three wires on one connection. S2T connects to B2T braid. The bottom of the antenna has S1B and S2B soldered together to complete the smaller loop. B1B and the braid of B2B are soldered together. I attach an SO-239 connector to the bottom by soldering the center conductor of B2B to the center of the connector and the braid of B2B to the connector's shell. The bottom now has two connections to the braid: one to leg B1B, the other to the shell of the connector. There's only one connection to the center conductor of B2B that goes to the SO-239 center pin.

Insulator Quality

A question arose concerning the dielectric quality of the tubing and pipe used for the insulating material. Antennas—being reciprocal devices—exhibit losses on a percentage basis, the percentage ratio being

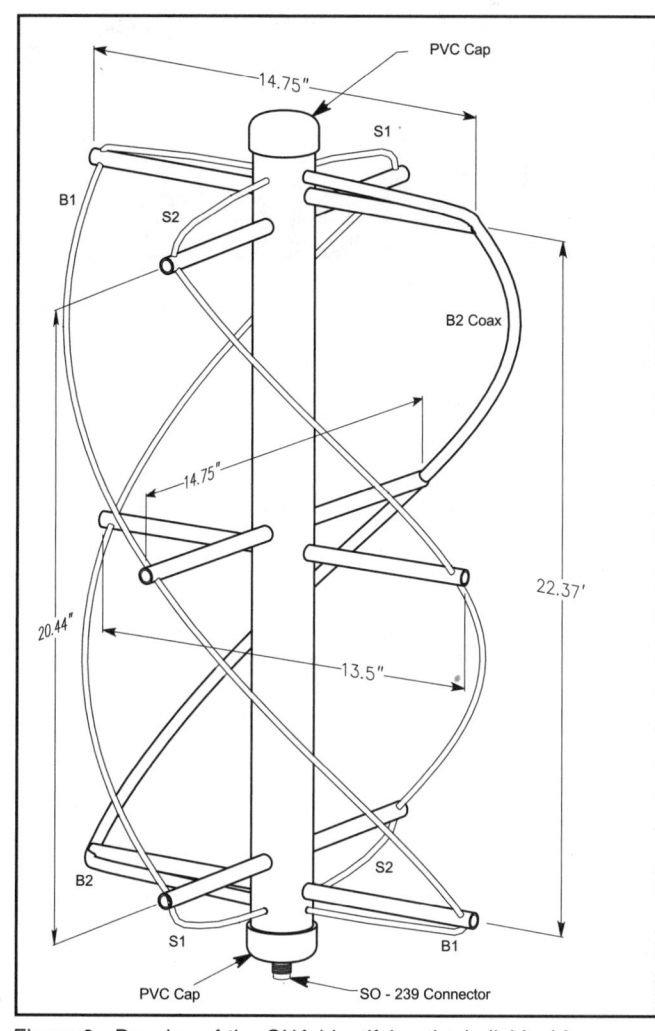

Figure 6—Drawing of the QHA identifying the individual legs; see text for an explanation.

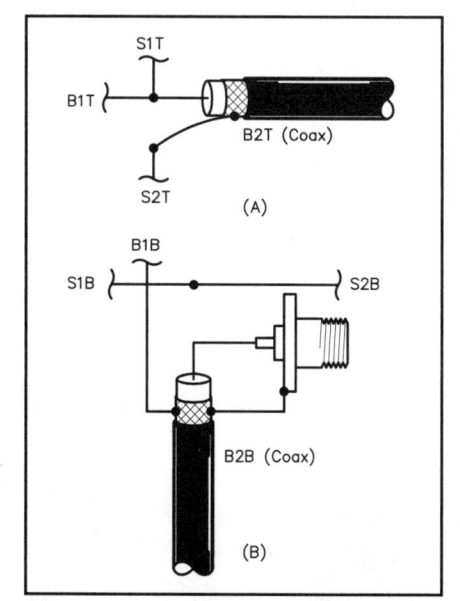

Figure 7—At A, element connections at the top of the antenna. B shows the connections at the bottom of the antenna. The identifiers are those shown in Figure 6 and explained in the text.

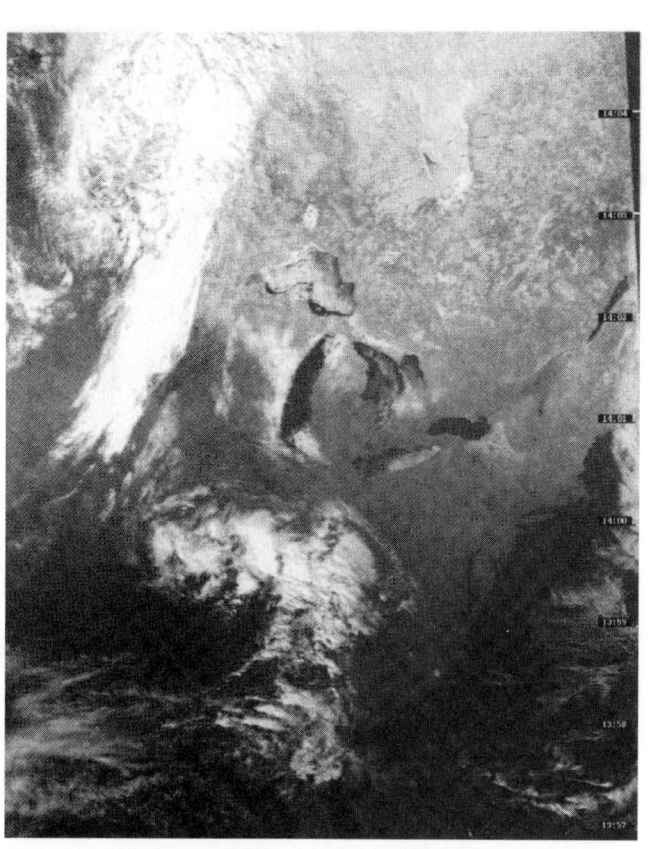

Figure 8—It's said that "The proof of the pudding is in the eating." To a weather-satellite tracker, clear, no-fade, no-noise pictures such as this one—compliments of W3KH's quadrifilar helix antenna—are delicious fare!

the same for transmit and receive. Although signal loss may not be as apparent on receive with a 2-µV signal as with a transmitted signal of 100 W (ie, it would be apparent if dielectric losses caused the PVC cross arms to melt!), signal loss could be a significant factor depending on the quality of the insulating material used in construction. As a test, I popped the pipe into the microwave and "nuked" it for one minute. The white PVC pipe and the tan CPVC tubing showed no significant heating, so I concluded that they're okay for use as insulating materials at 137.5 MHz or thereabouts.

The antennas cost me nothing because the scrap pieces of PVC pipe, tubing and connectors were on hand. Total price for all new materials—including the price of a suitable connector—should be in the neighborhood of $8 or less.

Results

I use a 70-foot section of RG-9 between the receiver and antenna, which is mounted about 12 feet above ground. As with the earlier antennas, I use a preamp in the shack. With AOS (acquisition of signal) on the first scheduled pass of NOAA-14, I was pleasantly surprised to receive the first of many fade-free passes from the weather satellites, including some spectacular pictures from the Russian Meteors! Although the design indicates a 3-dB beamwidth of 140°, an overhead pass provides useful data down to 10° above the horizon. (My location has a poor horizon, being located in a valley with hills in all directions but south.) I've also received almost-full-frame pictures of the West Coast and northern Mexico at a maximum elevation angle of only 12° at my location. (The 70-cm antenna works fine for PACSATs, although Doppler effect makes manual tracking difficult.) The weather-satellite antenna prototype worked better than expected and a number of copies built by others required no significant changes. The quadrifilar helix antenna is *definitely* a winner! And believe me, *it's easy to build!*

Acknowledgments

Thanks to Chris Van Lint, and Tom Loebl, WA1VTA, for supplying me with the necessary technical data to complete this project. A special thanks to Walt Maxwell, W2DU, for his review and technical evaluation and for sharing his technical expertise with the amateur satellite community.

Notes

[1] C. C. Kilgus, "Resonant Quadrafilar Helix," *IEEE Transactions on Antennas and Propagation,* Vol AP-17, May 1969, pp 349 to 351.
[2] M. Walter Maxwell, W2DU, *Reflections, Transmission Lines and Antennas,* (Newington: ARRL, 1990). [This book is now out of print.—Ed.]
[3] Randolph W. Brickner Jr and Herbert H. Rickert, "An S-Band Resonant Quadrifilar Antenna for Satellite Communication," RCA Corp, Astro-Electronics Division, Princeton, NJ 08540.

By Dick Stroud, W9SR

From *QST*, December 2002

Try Copper for 2 Meters —The Cu Loop

Want a rugged antenna with decent bandwidth for 2 meters? Try this "plumber's delight."

After the construction article for a 6 meter squalo appeared in *QST*,[1] I received several requests for a similar antenna for 2 meters. The unit shown here is made of standard ¾ inch thin-wall copper tubing available at nearly any hardware or plumbing supply store. This antenna is very sturdy, easily reproduced, performs well and can be built for about $10. It can be used in the vertical plane with FM repeaters and mobile stations or it can be flipped 90° for horizontal polarization, making it suitable for CW or SSB operation at the low end of the 2 meter band. The weight of the antenna, including mounting hardware, is about 2 pounds. The completed antenna can be seen in Figure 1, while Figure 2 shows the antenna rotated 90° for the polarization change.

Although construction is fairly easy, some basic plumbing techniques are required. The copper tubing and fittings shown here were purchased from a local ACE hardware supply store. For reference, the thin-wall elbows used to build this example were marked "EPC." Carefully cut the tubing to the lengths shown and assemble into the copper elbows. Make sure the sides are parallel and the two top sections are in line.

Clean the ends of the tube that will engage the elbows with steel wool and apply a paste soldering flux to this area. Solder the corners with a propane torch; keeping in mind that it takes time to get this much copper hot enough to properly flow the solder. Sufficient heat and good soldering technique will reward the builder with a professional looking joint. After soldering, remove any excess flux and polish the antenna using steel wool. Any extra solder buildup at the seams can be removed with a hand file, but proper soldering technique should avoid the need for this.

The bracket supporting the connector is made of 0.050 inch thick brass stock, as shown in Figure 3. The tubing is drilled and tapped at the center to accommodate the two 6-32 screws that attach the bracket. Sheet metal screws could also be used, if desired, but they should be stainless steel. The SO-239 coaxial connector is attached to the bracket with 4-40 stainless hardware. No aluminum is used in the signal path because dielectric action of the dissimilar metals could create electrical noise. The antenna mounting-bracket is not symmetrical, allowing additional clearance at the top, between the gamma match and the U-bolt.

The gamma-match shorting strap is made of 0.020-inch thin brass stock and is ½ inch wide. It is held in place with 6-32

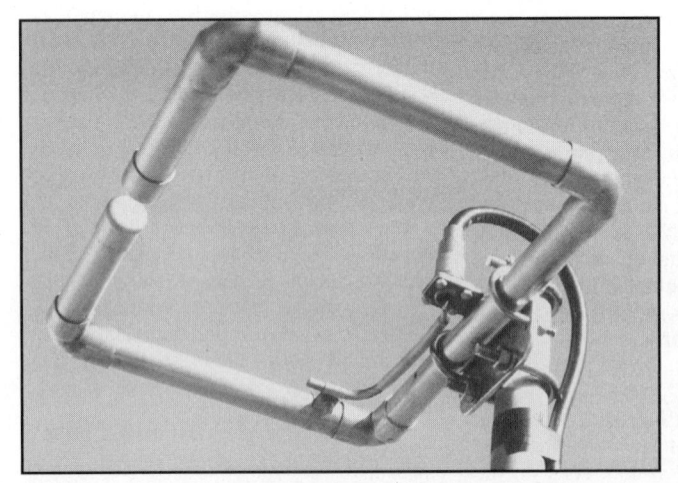

Figure 1—The Cu Loop mounted horizontally. Although not apparent from the photo, both the coax connector plug and socket are sealed to prevent moisture contamination.

Figure 2—(right) The Cu Loop mounted vertically. Note the gamma match strap. Make sure it is securely fastened to both the antenna element and the gamma rod, as the RF current is high at this junction.

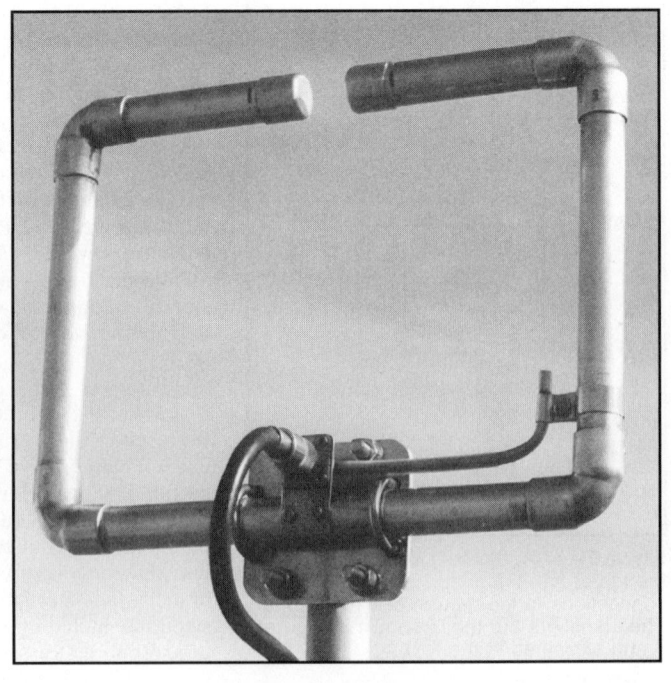

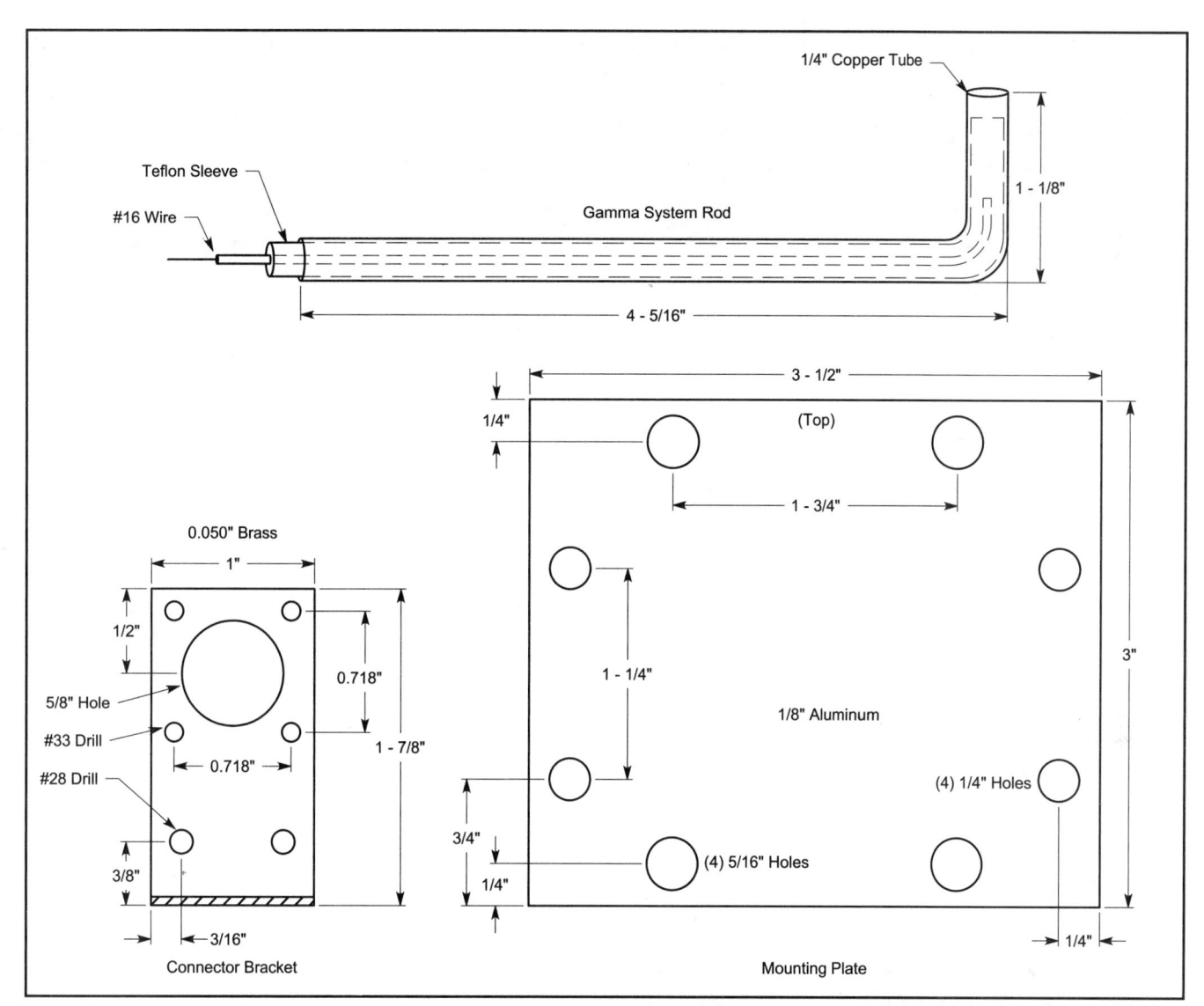

Figure 3—The gamma match, connector bracket and mounting plate details. The gamma rod to element connecting strap is not shown; it is described in the text. The hatched area of the connector bracket should be bent slightly to conform to the tubing radius.

stainless hardware. The gamma section is formed, as shown, from a 5⅝ inch length of 0.250 inch OD soft copper tubing, available at most hardware or plumbing supply stores. A 5½-inch length of 5/32 inch OD Teflon sleeving is inserted into the copper tube and a 5½ inch length of #16 vinyl insulated wire is placed inside this sleeving. The sleeving and wire are available from Mendelson Electronics.[2] See Figures 3 and 4 for the gamma match assembly details. The completed gamma match assembly should fit together snugly, and the wire end is then soldered to the coaxial connector center pin. The 0.250 inch copper tube and wire form the gamma match series capacitor, which measures approximately 10 pF.

With the gamma rod and coax connector in place, the two end caps can be installed. With both caps fully seated over their tubing ends, place a felt-tip pen ink reference mark on the tubes at the edge of the caps, as shown in Figure 5. Adjustment of both caps should then be made equidistant from these reference marks.

An RF analyzer (such as the Autek Research RF5 or MFJ 259/269) can be used to check and adjust the completed antenna. The antenna should be at least 6 feet above ground and be clear of surrounding objects during these adjustments and it should be supported on a mast or a test stand. The center frequency can be adjusted over a range of 138-149 MHz by sliding the end caps in and out on the antenna elements. Moving the caps closer together will lower the center resonant frequency by lengthening the antenna; this will also increase the capacitance across the loop.

Adjust the end caps equally to the desired operating frequency. On this example, the 2:1 SWR bandwidth was found to be about 4 MHz, at a center frequency of 147.00 MHz. If the dimensions have been followed carefully, the SWR should be very low. If necessary, the gamma strap can be moved slightly to minimize the reflected power and the wire length can be changed in small increments to bring the SWR to 1:1. The SWR can be adjusted to 1:1 anywhere in the band by adjustment of the end caps and the gamma section strap. After the adjustments are completed, both end caps should be secured to the antenna elements with #4 stainless steel sheet metal screws and the gamma strap tightened securely. The caps may be slotted to facilitate adjustment.

After testing and adjustment, a low-loss sealant should be placed over the back of the coax connector and at the ends of the gamma rod, to prevent moisture contamination. [Both the PL-259 connector and SO-239 socket are not waterproof. A preferred choice for continuous outdoor use would be a type N connector and socket. These are waterproof if properly installed.—*Ed.*] To preserve the finish and to keep the copper from oxidizing, the antenna can be coated with a clear protective finish, such as Krylon 1301.

This antenna has facilitated operation through many distant FM repeaters with gratifying results, while it was supported on a 20 foot test stand. It has also been used

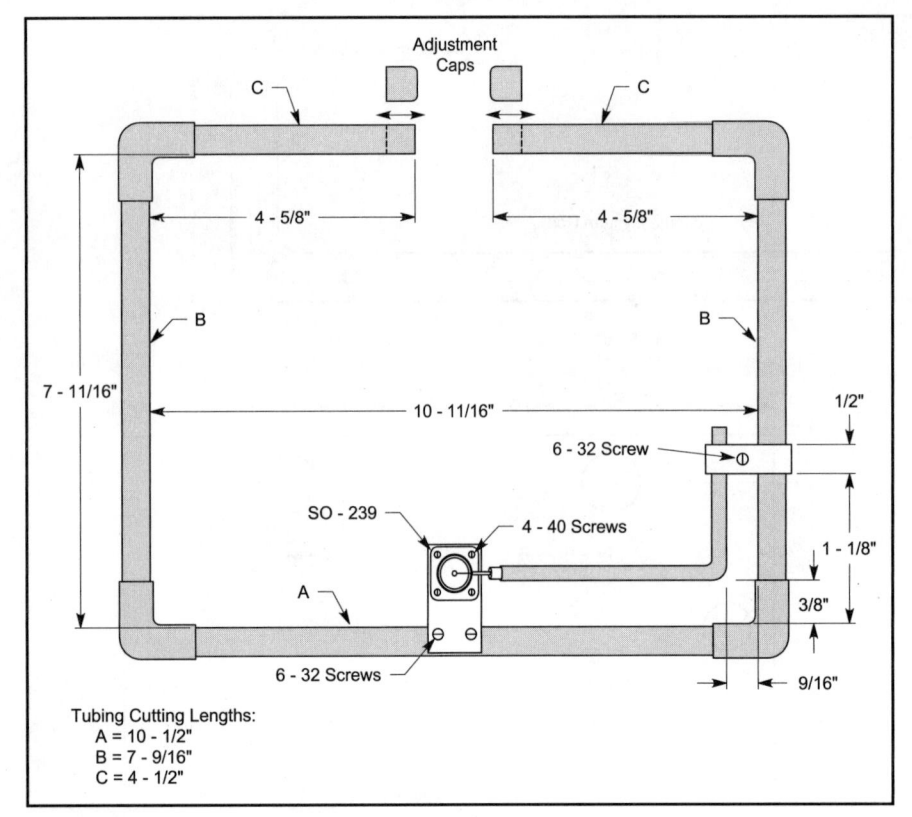

Figure 4—Construction details of the main antenna assembly. All hardware is stainless steel.

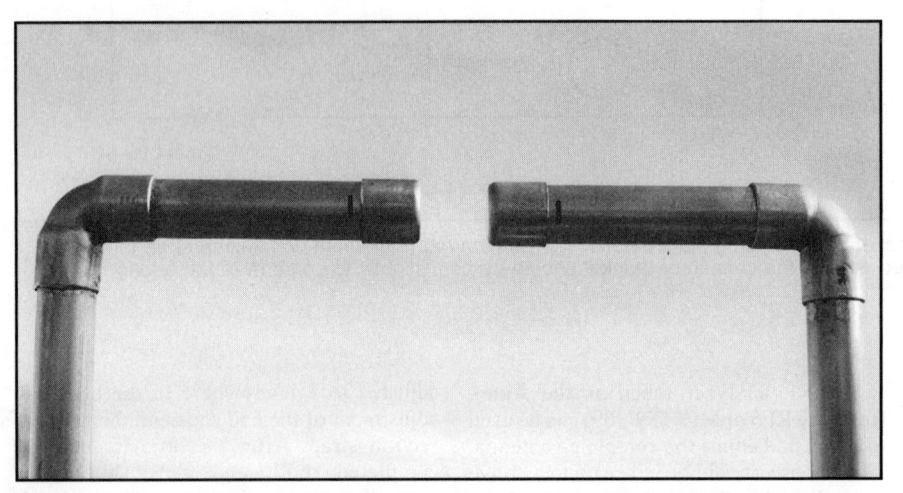

Figure 5—The adjustable tuning caps. Note the felt pen markings. Adjustments should be made equally from both end cap markings.

to make both CW and SSB contacts. The radiation pattern possesses a sharp null along a line drawn through the antenna's two sides. This characteristic, together with its small size and robust structure, makes it attractive for direction finding activities, in addition to general station use.

Notes
[1]"Six Meters from your Easy Chair," *QST*, Jan 2002, pp 33-34.
[2]Mendelson Electronics Co Inc, 340 E First St, Dayton, OH 45402.

By Jim Reynante, KD6GLF

From *QST*, January 1995

A Five-Element Quad Antenna for 2 Meters

If your station is located on the fringe of the repeater's coverage area, you've come to the right place. Why not build a 5-element quad antenna? It may be just the thing you and your radio need to provide a reliable communication link with the outside world.

My own location provides me with spotty coverage to the local repeater, and I knew that I needed a beam antenna to establish solid communication. I had seen designs for quad antennas in various books including *The ARRL Handbook*, yet I wondered how easily and inexpensively I could put one together for 2 meters. I chose low cost and simple construction as the main design goals. As a result, I ended up with an effective antenna that has the following features:

- A forward gain of at least 11 dBi
- An SWR of 2:1 or better throughout most of the 2-meter band
- Total construction time under two hours using simple hand tools
- Total cost less than $8 (depending on where you purchase your materials)

What, Me Build a Beam Antenna?

I made sure that building this antenna would be simple. In so doing, I practically guaranteed that I'd be able to complete the project! (I'm the last person that should be building an antenna. I'm typically "all thumbs" when it comes to construction practices.)

As you can see from the photograph, the antenna uses wood for the boom and dowels for the wire spreaders. It's oriented in the classic diamond configuration. The all-wood design allowed me to use simple hand tools for its construction, and the total cost for materials was just under $8 (see Table 1).

Construction

Before beginning construction, you must first determine the physical dimensions of each of the antenna's elements. The five elements of this quad antenna are the reflector, a driven element and three directors. The reflector is at one end of the boom, followed in order by the driven element and the first, second and third directors. Maximum radiation is along the line of the boom, in the direction of the third director.

Because the director element is arranged into a square loop one wavelength long, the actual length varies from the naturally resonant length. The lengths of the reflector, driven, and first director loop elements can be computed using the following formulas:

If you want an 11-dBi-gain beam antenna for less than $1 per dB, here it is!

Table 1
Materials List

Quantity	Description	Typical Cost
1	2×2×8-inch wood piece	$1.99
10	5/8-inch dowels, 30 inches long	$2.10
34 feet	#10 AWG bare copper wire	$3.06

$L_{Reflector} = 1071 / F_0$

$L_{Driven} = 998 / F_0$

$L_{Director1} = 973 / F_0$

where:

$L_{Reflector}$ = Length of the reflector element (in feet)

L_{Driven} = Length of the driven element (in feet)

$L_{Director1}$ = Length of the first director element (in feet)

F_0 = Center frequency (in MHz)

The lengths of the second and third director elements are determined by following a 3% series. In other words, the length of the second director is approximately 3% less than the first director, and the length of the third director is about 3% less than the second director. [The equations given in the text and the final dimensions in Table 2 are based on an optimized design for this antenna modeled at ARRL HQ with the *NEC2* computer program.—*Ed.*] Table 2 shows

144 MHz   2-43

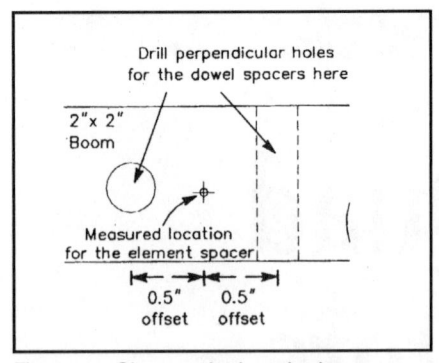

Figure 1—Close-up look at the boom shows mounting the hole offsets used for the dowel spacers.

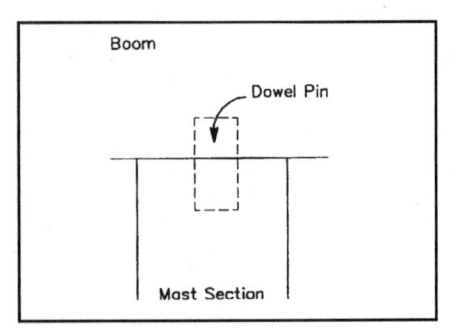

Figure 2—Use a 1-inch scrap piece from one of the dowels as a joint pin to secure the boom to the "shorty" mast section.

Table 2
Lengths and Spacing

Element	Element Length (inches)	Spreader Length (inches)	Spacing from end of boom
Reflector	88	31.25	2.5 inches
Driven	82	29	19.5 riches
1st Director	80	28.5	32.5 inches
2nd Director	78	27.75	48.5 inches
3rd Director	76	27	67.5 inches

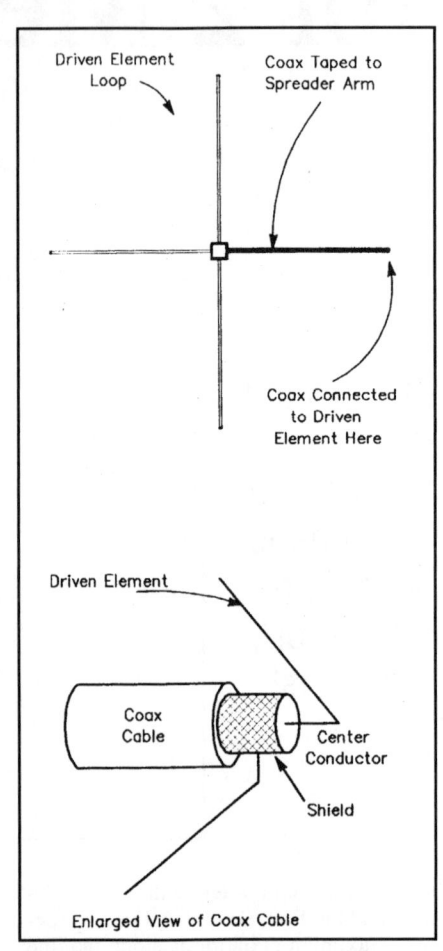

Figure 3—Feed the side corner of the driven element for vertical polarization.

the element lengths, spreader lengths, and element spacings.

With the element dimensions in hand, it's time to get started. Begin by preparing the boom. I decided to use wood for construction because of its low cost, wide availability and ease of use. When building antennas using wood, be sure to select pieces that are well-season and free of knots or damage.

Start with an 8-foot section of 2×2. Cut a 2-foot section from one end. (Don't discard it. It will be used to make a "shorty" mast section.) Starting from 2.5 inches from one end of the boom, carefully measure and mark the locations for each of the spreader arms. Each element requires two dowels to form the cross-arm assembly, so offset the hole locations by $^1/_2$ inch (see Figure 1). A drill press is ideal for drilling the holes, but an acceptable job can be done with a power hand drill. Be sure to mark which end is the reflector end and which end is the director.

Now carefully measure and mark the wood dowels that are used for the element spreaders. Use the dimensions in Table 2. You'll need two dowels cut to the same length for each element. After cutting the dowels, mount each pair into the appropriate hole locations on the boom, but don't glue them yet! Visually inspect the location of the spreaders on the boom. If everything's in place correctly, you should see the spreaders taper in length from the reflector end to the director end. Once you've verified the placement of the spreaders, you can secure them to the boom. Use a weather-proof glue or epoxy that is nonsoluble. Use a small saw or hobbyist's motor tool to carefully notch the ends of the dowels. These notches will be used to secure the wire elements.

Mount the boom to the remaining 2-foot length of 2×2. I used a simple butt-end joint reinforced by a small 1-inch long wood pin fashioned from a piece of scrap material from one of the dowels (Figure 2). Simply drill a small hole in the top of the mast about $^1/_2$-inch-long and the same diameter as your doweling. Drill a corresponding hole in the boom. Glue or epoxy the pin into the mast and boom to form the joint. For further reinforcement, you could also fashion gusset plates made from triangular pieces of $^1/_4$ inch plywood and attach them to both sides of the boom.

That completes the construction of the main antenna structure. Now, carefully measure and cut the wire used for each of the five elements. Do *not* use insulated wire! (Or at least remove the insulation.) Once the wire elements have been cut, it's time to mount them to the spreader arms to form the closed loops.

The reflector and director elements are strung around their respective element spreaders and held in place by soldering the ends together. At this point, you must decide whether to use horizontal or vertical polarization. The feedpoint of the driven element determines the polarization. Use corner feed on the side if you want vertical polarization (Figure 3), or attach the transmission line to the bottom corner for horizontal polarization (Figure 4). I opted for vertical polarization since I wanted to use it for FM and repeater work. If you want to try your hand at SSB and CW work, choose horizontal polarization. Attaching the transmission-line to the driven element may prove to be a little tricky, so an extra pair of hands may help. I found it useful to tape the transmission-line spreader arm to hold it in place while I soldered it to the driven element.

I used RG-8X for my setup since I only had a short run (less than 20 feet) to my shack. For installations with longer runs, I recommend using standard RG-8 coax or better to deliver every dB you can to the antenna. Apply a silicone or similar sealer to the exposed end of the coax to prevent the possibility of moisture seeping into the line. The wood surfaces should also be varnished to help protect them from the weather.

This five-element quad is a real performer. If you build the antenna as shown, the computer model predicts 11 dBi gain and a front-to-back ratio of 20 dB. ARRL Lab measurements confirmed the computer-predicted antenna pattern.

You may wish to read further about tuning adjustments in *The ARRL Handbook*. It describes how to check the antenna with a field-strength meter, or by using a separate receiver in conjunction with a dipole antenna.

Field Results

After connecting the quad to my radio, I rechecked the SWR. Figure 5 shows the

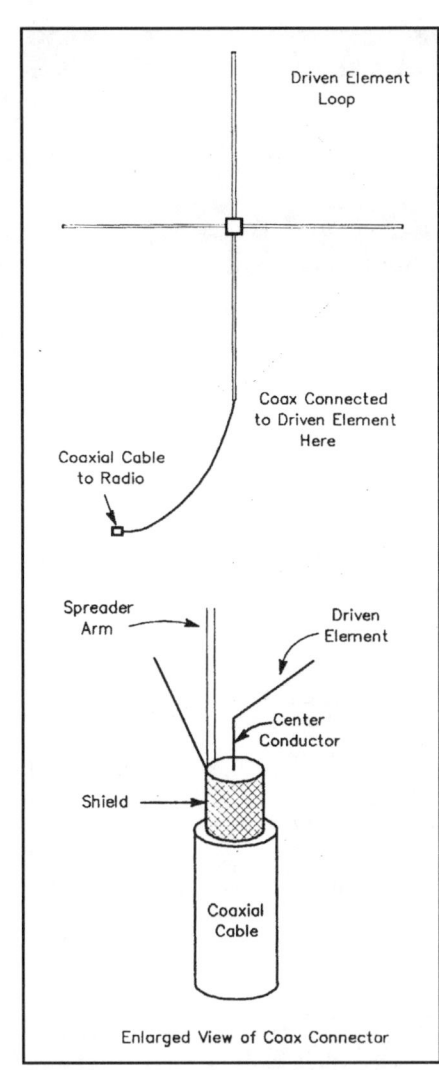

Figure 4—Feed the bottom corner of the driven element for horizontal polarization.

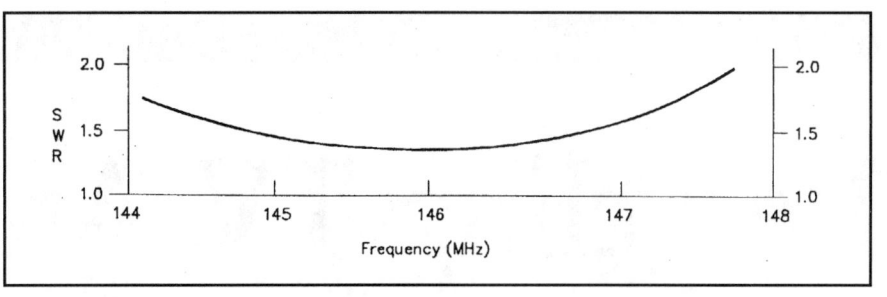

Figure 5—Here are the SWR readings the author obtained. When the antenna was constructed and tested by the ARRL Lab, even better SWR results were obtained.

SWR readings I obtained throughout the 2-meter band. It's interesting to note that the lowest SWR occurs just below the center frequency of 146 MHz. This is due to the fact that I cut the elements a little long to allow for trimming. I suggest you do the same. Shortening the wires is simply a matter of cutting off the extra length and then deepening the notches in the dowels.

After I had established contact with a friend on the local repeater, we decided to try the quad by switching to simplex operation. (In the past, he and I were unable to communicate directly.) I punched in 146.535 MHz on my radio and rotated the quad toward his station. I nervously called him. After waiting for what seemed like an eternity, I heard his answer! We proceeded to carry on a casual ragchew discussing antennas, current events and so on without tying up the local repeater. (The joys of simplex!) Before signing off, I slowly rotated my quad while he talked in order to get an audible indication of the quad antenna's side and back directivity. His signal faded as expected, and then returned as the quad once again pointed in his direction.

When using the antenna connected to my handheld transceiver with only 1.5 W of output power, I'm able to consistently access repeaters 40 to 50 miles away. Everyone says my signal is full quieting. That's not too bad, considering I had been unable to even hear those repeaters using my quarter-wavelength ground-plane antenna—which is mounted about 10 feet higher than the quad.

Summary

I've been enjoying this antenna for the past few months without any problems whatsoever. I've logged a couple of hundred contacts, both simplex and via repeaters. I've also used it with equal success for operating packet.

Don't be disheartened if you happen to live in a community with difficult antenna restrictions. In addition to being inexpensive and simple to construct, this quad antenna is also compact enough to be placed in an attic!

By Ronald Lumachi, WB2CQM
From *QST*, November 1996

Recycling TV Antennas for 2-Meter Use

"Green" that aluminum into something useful right in your own backyard!

Just as quickly as cable TV companies wire up neighborhoods for CATV service, home owners remove TV antennas, masts and dangling wires. The quality of reception and the overall appearance of the old homestead are immediately improved even before the ladder is hung in the garage and the scrap aluminum components are bent, bundled, and stacked curbside for the next trash pickup.

Wait a minute!

Are you sure you want to bid farewell to that mass of metal? These slowly disappearing aluminum skyline silhouettes are a gold mine of salvageable antenna construction material. When modified by a resourceful radio amateur, they'll become—with almost no cash outlay—a respectably performing Yagi antenna for the 2 meter band.

What's of Value In that Pile of "Junk"?

You may be surprised at what can be salvaged from discarded TV antennas and

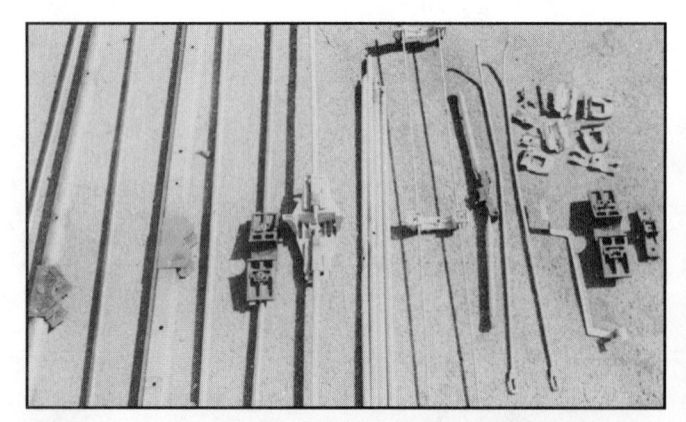

Figure 1—A gold mine of antenna components after complete disassembly. Note the two boom/mast clamps to the left. The two types of insulated elements are positioned at the center of the cluster. Either can be used in the project. **U** bolts, conspicuously absent, were rusted. RadioShack replacements were used.

Figure 2—Drill out the rivets to remove the TV-antenna pieces. Notice the element-mounting hardware to the left; it grounds the element directly to the boom. Some of these clamps can be used to mount the directors. The phasing stubs visible below the boom can be discarded.

2-46 Chapter 2

used for your VHF-antenna construction projects (see Figure 1). Obviously, the boom has value once the elements are removed. (A drill and an assortment of twist bits make short work of the rivets used in the original assembly.) If two similar TV antennas are available, making a stacked array is a viable option. Or Long John high-gain Yagis can be constructed if two booms are butted and clamped together.

You'll find a variety of clamps used to secure the elements to the boom. Most attach the elements directly to the boom (see Figure 2), but at least one of the elements—the driven element—is insulated from the boom by a nonconductive support. Set one of these insulated elements aside to be used for the 2 meter antenna's driven element (discussed later). Driven elements can be used as directors, but their individual sections must be wired together (bridged) and grounded to the boom (see Figure 3).

The TV antenna's elements (rod stock, if you're lucky [*really* lucky!—*Ed.*]) can be cut to resonate on 2 meters. In addition, there are plastic boom-end caps and the boom-to-mast connecting hardware. If you're fortunate, the boom-mounting U bolt can be removed and reused if care and some Liquid Wrench are applied in the process. If the U bolt is too rusty, get a replacement from RadioShack or one of the many other outlets.

Reconstructing the Antenna for 2-Meter Operation

For this project, we'll build a five-element Yagi as shown in Figure 4. If the TV antenna's reflector-mounting hardware is in good condition, leave that element attached to the boom. This element (the longest of them all) is usually grounded at its center point. Hacksaw the boom to a length of 5 feet 5½ inches and replace its end caps. The caps cut down on the wind load and should not be overlooked. [They also help decrease wind noise.—*Ed.*]

If you want to orient your soon-to-be 2-meter antenna for vertical polarization, move the boom-to-mast clamp 90° from its original mounting position. At a point slightly off the boom's center position, drill two ¼-inch-diameter holes through the boom using the U bolt as a guide. Make certain that the boom-to-mast clamp won't interfere with the mounting of the first director (the one closest to the driven element). If necessary, move the clamp position slightly for clearance. Its position (within reason) isn't critical for balance or rigidity.

For this antenna, we're using an element spacing of about 0.2 λ. After trimming the remaining directors to length, attach them to the boom, spaced accordingly. Use bolts, self-tapping screws, or pop-rivets. To fabricate elements of sufficient length, it may be necessary to butt two shorter pieces and couple them with a 3 or 4-inch-long sleeve.[1] Such sleeves can be fabricated from short lengths of element tubing. If the element is a continuous tube, slit it lengthwise. If the element is a rolled tube (split down its length), carefully fold it into itself (see Figure 5) using a vise or pliers to obtain the correct OD. Secure the element to the boom at the butt joint with the TV hardware.

Driven-Element Construction

We'll use simple inductive coupling to drive the array. First, disassemble all the components from the original TV antenna's driven element. Keep the insulator and the two element sections. The remaining parts of the subassembly—including the phasing stubs (the crisscrossed wires on the antenna; see Figure 2 and the lead photo)—can be discarded. Drill out the rivets securing the elements to the insulator. Reattach this assembly to the insulator using the flat areas of the clamp. For this element, the element-to-boom fastenings must not

Figure 3—A director attached to the boom using an insulated mount. The two element sections have been bridged with short lengths of wire, solder lugs and pop rivets, and grounded to the boom with a self-tapping screw. Two of the antenna's original rivets and remnants of the phasing stubs are visible at the base of each element.

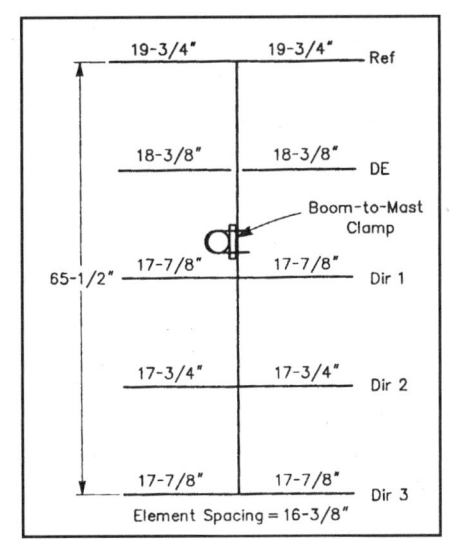

Figure 4—Dimensions of the five-element 2-meter Yagi. The stub is inside the driven element (see text).

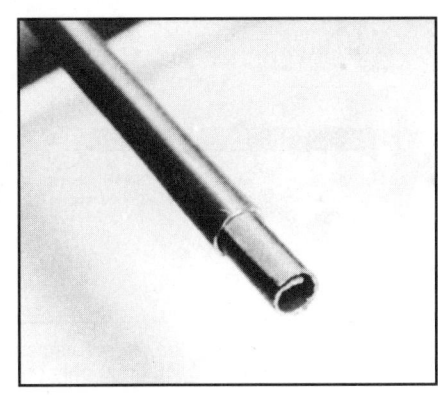

Figure 5—Close-up of a short length of scrap (rolled) TV element folded in on itself and used as an internal butt-joint splice to increase the length of too-short elements. Mounting hardware is placed over the joint, drilled through, and attached to the boom with a self-tapping screw.

Figure 6—A close-up of a fully modified 2-meter driven element. The two top-mounted rivets that originally secured the element to the insulator have been removed. Two new pop rivets at the side of each element section affix the element-mounting hardware to the insulator. The inductive-stub ends are just visible exiting each element section. An L-shaped bracket secures the optional SO-239 connector to the boom. A short length of RG-58 coaxial cable connects the matching stub to the SO-239.

---

[1]Considering the weathered, oxidized and corroded condition of most (if not all!) of the metal you're likely to encounter when using junked TV antennas, think about ensuring good contact at metallic joints during reassembly. Buff the mating metal joints (not all the aluminum) and apply an antioxidant. For a discussion on fighting corrosion and some hints as to which products to use, see Scott Roleson, KC7CJ, "Fighting Antenna Corrosion," *QST*, Apr 1993, pp 24-26—*Ed.*

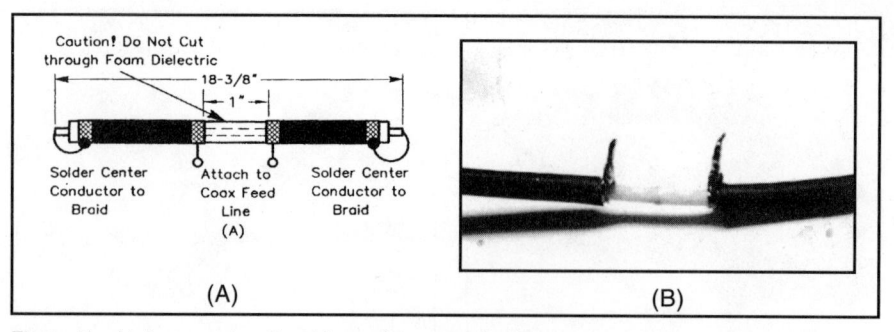

Figure 7—At A, preparing the driven-element stub. Use a razor blade or sharp knife to carefully cut through the coax braid only (*not* the dielectric) at the stub center. Unravel the braid away from the center point. At B, a close-up of the stub. The shield is carefully cut through on the center line. The twisted and tinned braid leads are connected to the coaxial feed line.

obstruct access to the interior of the driven elements at the midpoint (see Figure 6). You need to keep this path clear in order to allow passage of the matching stub.

To construct the stub (see Figure 7), cut a length of RG-58 cable to $18^3/_8$ inches. At each end of the cable section, remove the outer covering and dielectric, and solder the shield braid to the center conductor. Cover the shorted ends of the stub with heat-shrink tubing or electrical tape to keep them from shorting to the driven element. At the stub's center, remove about 1 inch of outer covering. Use a razor blade or sharp knife to carefully cut through only the braid (not the dielectric) at the stub midpoint (see Figure 7B). Unravel the braid away from the center point. Twist and tin the braid making certain not to melt the dielectric. Later on, these two braid leads will be connected to the coaxial feed line.

Attach the driven-element assembly to the boom $16^3/_8$ inches from the reflector. Slide the taped ends of the stub into the driven element. Center the two tinned braid leads over the boom. Solder the feed line directly to the stub's braid leads. Or install an SO-239 or BNC connector as shown in Figure 6, connecting the stub's leads to it. Once you've made the connections, protect this area from the weather with a layer of noncorrosive silicone sealant. If you've constructed the antenna as described, you'll find the inductive coupling network (stub) provides a low SWR over the entire 2 meter band, so you shouldn't have to do any tweaking. Crimp shut the ends of each element to reduce wind load, vibration and noise. Finally, attach the array to your mast and rotator and you're on your way!

Summary

To ham homebrewers, the buzzword "recycling" sounds like déjà vu all over again. I can remember when hams were continually on the prowl for discarded tube-type TV chassis for their wealth of electrical components and the heavy duty transformers they contained. Before the trash pickup got to them, they were unceremoniously carted home and picked clean of the hundreds of parts that subsequently filled the junkbox. Only then—like the carcass of a Thanksgiving turkey—was the gutted chassis, with its cut wires and empty tube sockets, reluctantly relegated to the junk heap.

Although this 2 meter antenna has been resurrected from a pile of scrap aluminum components, it's no Frankenstein monster. We've recycled a valuable resource, completed a worthwhile project and improved signals by many decibels with a minimal cash outlay. Sounds like an undeniably good deal to me!

By R. J. Decesari, WA9GDZ/6

From *QST*, September 1980

A Portable Quad for 2 Meters

Backpacking, boating or mountaintopping? Invest an afternoon's work and pack this novel directional gain antenna on your next expedition.

Last year, while I was "hilltopping" in the San Diego area with my 2 meter FM transceiver, a band opening occurred in which stations from Los Angeles, Santa Barbara and further points north were copied on simplex frequencies. Establishing solid communications with the built-in quarter-wave whip antenna and 1-watt power of the transceiver (with weakening batteries) was rather difficult, even with the opening. Because of my intense desire to communicate with these DX stations, a need for either a directional gain antenna or a power amplifier was established. Since I didn't particularly desire toting and charging additional batteries for an amplifier, I set this concept aside. I then took a closer look at improving the antenna. This novel portable antenna configuration evolved from many hours of thinking and tinkering in my workshop.

Initial efforts to design a collapsible antenna centered on a conventional four-element Yagi configuration. Several models of the Yagi, whose elements all opened simultaneously, proved to be a nightmare in bell cranks and lever arms. From this attempt, I decided that all the elements should still be attached to a main boom, but the operator would open the elements individually during antenna setup, thus eliminating the push rods and cranks. The Yagi design, with the elements folding on top of each other to minimize space, was still rather large considering element spacing and other required mechanical appendages and dimensions. At about this time, I happened to spot a big 20 meter quad while driving to work and immediately started to

Figure 1 — The basic portable quad assembly. The author used an element spacing of 16 in. (406 mm) so that the quad spacers would fold neatly between the hubs.

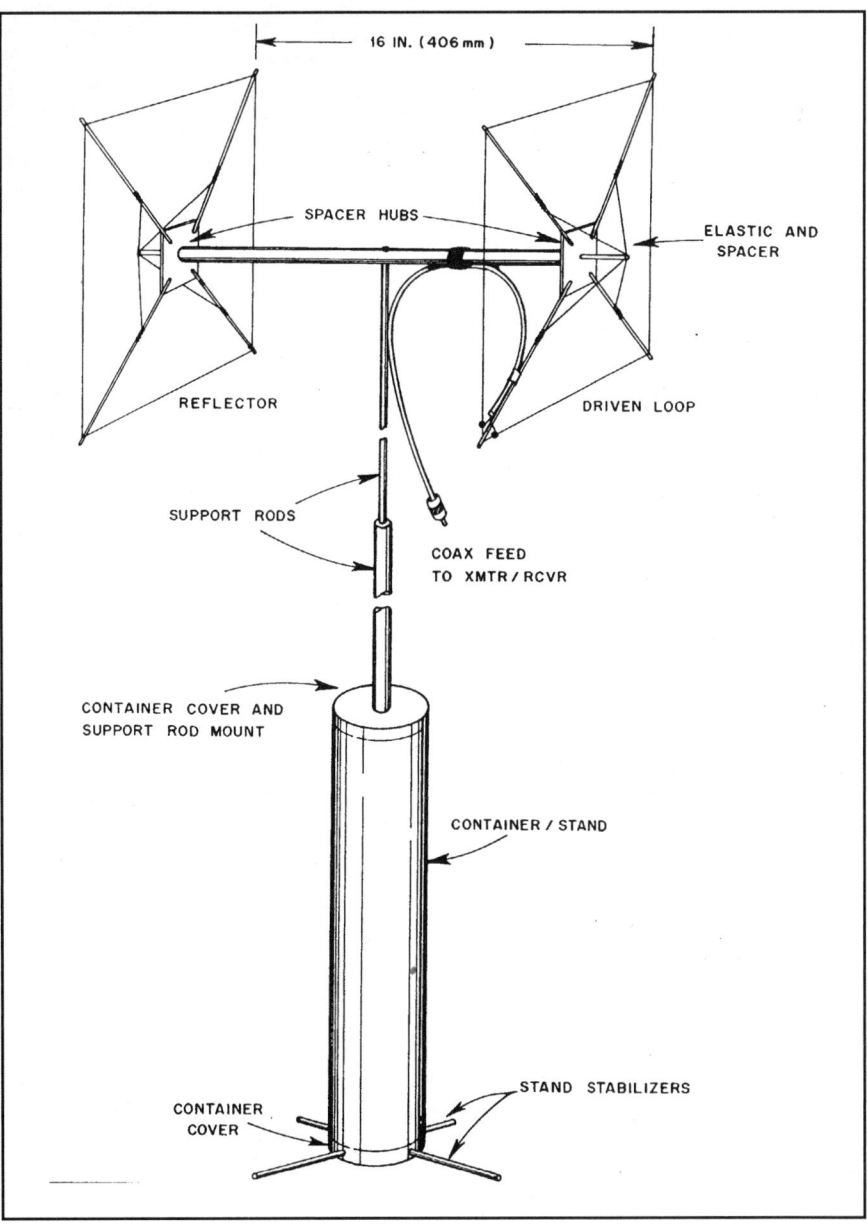

144 MHz 2-49

ponder the possibilities of using a quad for the intended portable antenna.

With only two elements, the quad provides an excellent front-to-back ratio, as well as about 6 dB of forward gain. With a two-element quad, the element spacing for optimum reflector performance is between 0.15 λ and 0.2 λ. That works out to about 12 and 16 inches (305 and 406 mm) at 2 meters. Not a bad overall size for a 2 meter antenna! Now the problem was how to support the square loops. A quick lesson in geometry revealed that if an "X" configuration of spacers were used to support 144-MHz loops, then each leg of the "X" would also be about 16 inches! All that was left to do was design a center hub that would allow the spacers to fold to the longitudinal axis of the boom and the basic problem would be solved. Consequently, the garage workshop was put into overtime service and the preliminary model of the brainchild was fabricated.

A Quad is Born

Figures 1 and 2 show the basic portable quad. Both driven and reflector elements fold back on top of each other, resulting in a structure about 17 inches (432 mm) long. The wire loop elements may be held in place around the boom with an elastic band. To support the antenna once it has been erected, the container is used as a stand. To provide more stability, four small removable struts slip into holes in the base of the container. Both the support rods and struts fit inside the container when the antenna is disassembled.

I have used two different methods of keeping the quad spacers erect. Both methods are successful. Figure 3 shows the quad spacers held open by spring-steel clips. Each clip is fabricated from an ordinary paper binder with a hole drilled in it to allow it to be attached to the quad spacer. The clip is compressed and slid down the quad spacer until it engages the hub. This provides a rigid mechanical support to hold the spacer open when in use as well as allowing it to pivot back for storage in the container. Figure 4 shows a slightly different method: A mechanical stop is machined into the hub, and elastic bands are used to hold the spacers erect. The bands are attached to an additional strut to hold the spacers open. When not in use, the strut pulls out and sits across the hub, and the spacers can be folded back. Details of each method are shown in Figure 5.

The clip and hub assembly is possibly easier for the home builder to fabricate, with the exception of drilling the hole in the spring steel. A high-speed-steel or carbide-tipped drill set is required, since the spring steel is an extremely tough and brittle material. Care must be taken when drilling the holes since the clip material will tend to crack. It is recommended that the builder start with a small-diameter drill and proceed to sequentially larger drill diameters until the final diameter is reached. The clip should be expanded and fitted over a

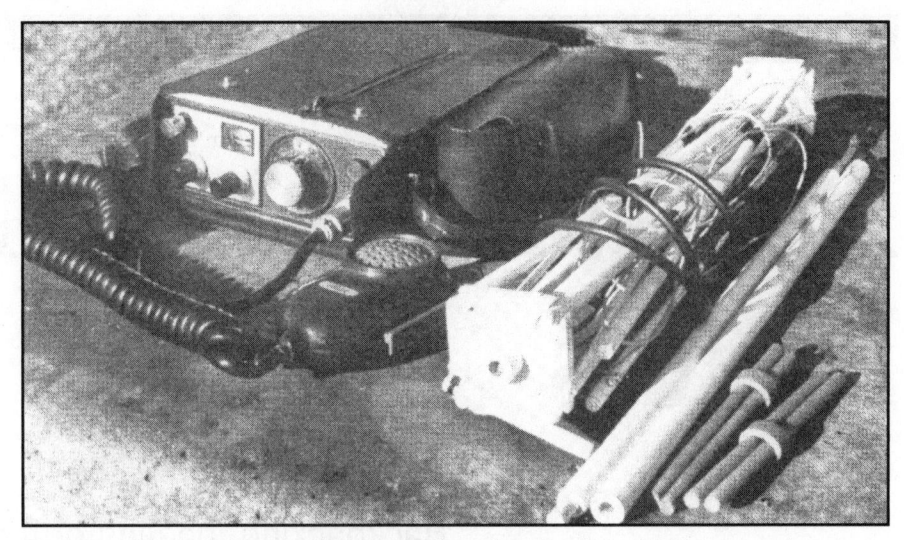

Figure 2 — The portable quad in stow configuration. Two long dowels are used as support rods. Four smaller dowels are used to stabilize the container.

Figure 3 — Paper-binder spring clips are used in this version of the quad to hold the spacers erect.

Figure 4 — This version of the portable quad uses mechanical stops machined into the hub; elastic bands hold the spacers open.

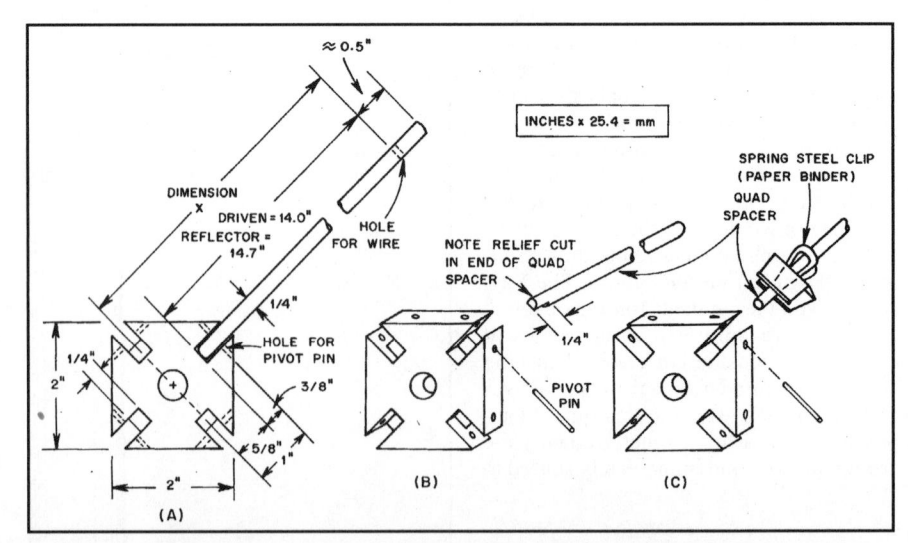

Figure 5 — A detail of the spacer hub with spacer lengths for the director and reflector is shown at A. The hub is made from 1/4 inch (6.4 mm) plastic or hardwood material. The center-hole diameter can be whatever is necessary to match the diameter of your boom. The version of the hub with mechanical stops and elastic bands is shown at B. At C is the spacer hub version using spring clips.

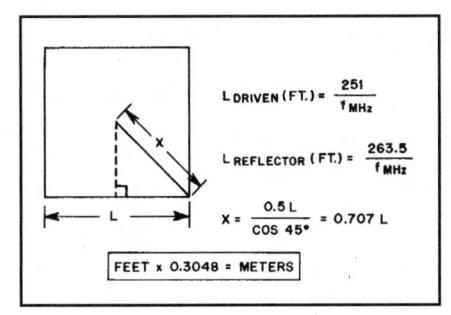

Figure 6 — Quad loop dimensions. Dimension X is the distance from the center of the hub to the hole drilled in each spacer for the loop wire. At 146 MHz, dimension X for the driven element is 1.216 feet (14.6 inches), and dimension X for the reflector is 1.276 feet (15.3 inches).

$$L_{DRIVEN}(FT.) = \frac{251}{f_{MHz}}$$

$$L_{REFLECTOR}(FT.) = \frac{263.5}{f_{MHz}}$$

$$X = \frac{0.5 L}{\cos 45°} = 0.707 L$$

FEET × 0.3048 = METERS

1/4 inch (6.4 mm) piece of wood to be used as a drilling back. Use of a light oil is recommended to keep the drill tip cool.

Building Materials

The portable quad antenna may be fabricated from any one of several plastic or wood materials. The most inexpensive method is to use wood doweling, available at most hardware stores. Wood is inexpensive and easily worked with hand tools; 1/4 inch (6.4 mm) doweling may be used for the quad spacers, and 3/8- or 1/2 inch (9.5 or 12.7 mm) doweling may be used for the boom and support elements. A hardwood is recommended for the hub assembly, since a softwood may tend to crack along its grain if the hub is impacted or dropped. Plastics will also work well, but the cost will rise sharply if the material is purchased from a supplier. Plexiglas is an excellent candidate for the hub. Using a router and hand tools, I manufactured a set of Plexiglas hubs with no difficulty. Fiberglass or phenolic rods are also excellent for the quad elements and support.

The loops were made with no. 18 AWG insulated stranded copper wire, although enameled wire may also be used. If no insulation is used on the wire and wood doweling is used for the spacers, a coat of spar varnish in and around the spacer hole through which the wire runs is recommended. The loop wire terminates at one element by attaching to heavy-gauge copper-wire posts inserted into tightly fitting holes in the element. For the driven element, two posts are used to allow the RG-58/U feed-line braid and center conductor to be attached. A single post is used on the reflector to complete the loop circuitry.

The first model of this antenna had a tuning stub attached to the reflector loop. This allowed a certain degree of reflector tuning to maximize its performance. However, I discovered a computer maximization of quad loop and spacing dimensions.[1] This data was used in my subsequent 2 meter quad designs, and has simplified the antenna by eliminating the need for a reflector tuning stub. Figure 6 shows quad dimensions derived from this data. The quads described in this article have been designed for 146 MHz, but the basic loop size equations will allow the builder to construct a model to any desired frequency in the 2 meter band to maximize results.

The storage container was made from a heavy cardboard tube originally used to store roll paper. Any rigid cylindrical housing of the proper dimensions may be used. Two wood end pieces were fabricated to cap the cardboard cylinder. The bottom end piece is cemented in place and has four holes drilled at 90° angles around the circumference. These holes hold 4-inch (102 mm) struts, which provide additional support when the antenna is erected. The top end piece is snug fitting and removable. It is of sufficient thickness (about 5/8 inch or 16 mm) to provide sufficient support for the antenna-supporting elements. A mounting hole for the supporting elements is drilled in the center of the top end piece. This hole is drilled only about three-quarters of the way through the end piece and should provide a snug fit for the antenna support. One or more antenna support elements may be used, depending on the height the builder wishes to have. Keep in mind, however, that the structure

[1]"Optimum gain element spacing found for the quad antenna," *World Radio News*, March 1978.

Figure 7 — A vertically polarized portable quad. The feed point is at the extreme left of the photograph.

The author with the fully erected portable quad antenna. The bottom stand is also used as a storage container.

will be more prone to blow over, the higher above the ground it gets! Doweling and snug-fitting holes are used to mate the support elements and the antenna boom.

Polarization and Performance

The antennas shown in Figures 1 through 4 all have 45° diagonal polarization. This is a compromise between vertical and horizontal polarization that allows both FM and SSB/CW (which is usually horizontally polarized) to be worked on 2 meters. Figure 7 shows another version of the antenna, built for vertical polarization. Although analytical antenna-pattern and gain tests have not been conducted, the portable quad displays an excellent front-to-back ratio as well as gain. The antenna has been used in the field with very satisfying results. The best example of the performance of the antenna was demonstrated by comparison to a 5/8 wave whip antenna. In this demonstration, the 5/8 wave whip was placed on a table top inside the ham shack and excited with 15 watts. From a location in San Diego, the 5/8 wave whip was unable to trigger any of the Los Angeles repeaters about 150 miles to the north. With the portable quad sitting on the same table, full-quieting access was gained to the Los Angeles repeaters.

This antenna design provides a compact package for a directional-gain antenna ideally suited for portable operation. Furthermore, it can be built from readily available and inexpensive materials. I would like to thank my father-in-law for his encouragement and my wife Sue for her patience and indulgence.

Chapter 3
432 MHz

Optimum Design for 432 MHz–Parts 1 and 2   Steve Powlishen, K1FO   3-1

A 432-Yagi for $9   Dave Guimont, WB6LLO   3-13

By Steve Powlishen, K1FO

From *QST*, December 1987

An Optimum Design for 432-MHz Yagis

Part 1: What's involved in designing a high-performance array? This month: some of the parameters that must be considered in an antenna design.

The latest rage in 432-MHz Yagi design seems to be extremely long antennas—more than 10 wavelengths. I spent nearly two years perfecting two different 10.5-wavelength designs (24-foot boom length) and presented the results of those efforts in 1986 at the major VHF/UHF conferences.[1] Frank Potts, NC1I, successfully used one of the 10.5-wavelength designs in a 16-Yagi earth-moon-earth (EME) array. Frank's success encouraged me to plan a 26.0-dBd-gain array using eight of those Yagis. The antennas would be stacked two wide by four high and located 80 feet high for tropo and EME use. After thoroughly researching this planned array, however, I came up with a different solution. This article describes my efforts.

Will It Stay Up?

An often overlooked antenna design consideration is wind load. Wind load is the force put on a structure by wind blowing against it. I live in an area where ice loading combined with moderate winds can quickly destroy a poorly engineered antenna system. With this in mind, I analyzed possible antenna configurations for gain versus wind load.

Calculations showed that the eight-Yagi array would cause the collapse of my present tower the first time winds exceeded 40 mph! The eight long Yagis, stacking frame and cables exhibit a total wind load area of nearly 40 sq ft. Since the Yagis were to be elevated for EME, the array had to be centered more than 14 feet above the top guy wires. I calculated the bending moment for this configuration in an 80 mph wind. The resultant force of almost 17,000 footpounds is nearly the level that would collapse an 18-inch-face commercial steel tower and three times that which would destroy a 12-inch-face steel tower. These sobering figures encouraged me to find another way to construct a high-gain 432-MHz array.

Next, I examined a plan that used 16 moderately sized Yagis. Using low-wind load, lightweight antennas (similar in size and mechanical construction to the popular K2RIW 19-element design), a 16-Yagi array has a wind load area of 32 sq ft. The antenna booms are much shorter, so this array has to be mounted only nine feet above the top tower guys. The resultant bending moment on the tower for the array of 16 shorter Yagis is a manageable 8600 foot-pounds—half that of the array of eight long Yagis. I anticipated that the 16-Yagi array would have 27.0 dBd gain. In terms of gain versus wind load, the array of shorter antennas seemed to be a better approach. The only penalty would be a more complicated feed system.

As an interim step, I decided on a 25.7-dBd-gain array of 12 shorter Yagis. The 12-Yagi array has a wind load area of 24 sq ft and a bending moment of 6500 foot pounds. My over-guyed tower could handle this antenna.

Yagi Development

Once I decided on the array configuration, the next step was to choose an antenna design. The 19-element K2RIW Yagi (RIW 19) is enormously popular in North America because of its
- light weight
- low wind load
- clean pattern (except for rear lobe)
- self-supporting boom (no braces required)
- good wet and dry weather performance.

I had been using arrays of RIW 19s for several years with good success. I had learned enough from working with Yagis extensively over the past few years, however, to convince me that a much better design could be had within the same approximate wind load as the RIW Yagi. In March 1986, I started work on a new moderate sized Yagi, one that I hoped would become a replacement for the RIW design. The design criteria included
- low wind load (<1 sq ft)
- light weight (<4 lbs)
- no boom support
- gain about 1 dB better than the RIW 19
- improved pattern compared to the RIW 19[2]:
- E-plane sidelobes –17 dB or better
- H-plane sidelobes –16 dB or better
- H-plane minor lobes substantially better
- rear lobe –20 dB or better (5 dB improvement)
- lobes surrounding the rear lobe –25 dB

Much of the design and analysis work was done using MININEC, a microcomputer-based antenna analysis program. Modeling antennas on the computer makes it possible to try many designs without drilling a single boom. I do not have sufficient computer power, or a sophisticated enough Yagi analysis program, to optimize element spacing and element lengths simultaneously. I started with a spacing pattern based upon knowledge and experience and used the computer to optimize element lengths. Several possibilities came to mind.

Modifying the RIW 19. Much computer time was spent on this approach because I had a significant investment in RIW 19 Yagis (12 of them to be exact). Although I found I could get more gain out of an RIW 19 by making a longer center boom section and using a single reflector, the RIW design was compromised by change. The target gain could not be reached with a reasonable boom length while keeping a clean pattern.

Using the DL6WU design. The antenna design by Gunter Hoch, DL6WU, is an excellent performer. Its flexible design (it can be made 2 to 14 wavelengths long) requires a trade-off, though: The Yagi will not be optimized for any given boom length. Using the DL6WU element spacing yielded a 20-element Yagi on a 13-ft, 8-in boom. Computer analysis indicated that this antenna would not meet my gain target. If an additional director was added, the boom would be 14 ft, 7 in long. I don't feel comfortable using a self-supporting small-diameter boom of this length. I had previously optimized the DL6WU director lengths for a 31-element, 24-ft Yagi. Reducing this antenna to a 20-element, 13-ft, 8-in Yagi did not give good results; that optimization had negated the variable length design feature.

Starting from scratch. In this day and age, starting from scratch is almost like reinventing the wheel, and it soon became apparent that it would take a major effort to design from scratch an antenna that would outperform the modified DL6WU design. My objective was to create a better EME array; it was not merely a theoretical exercise.

The W1EJ Designs

Fortunately, I found someone who had *already* reinvented the wheel! Tom Kirby, W1EJ, had spent several years working on computer-optimized Yagi designs. Tom found that each Yagi size needs its own set of spacings to achieve the best combination of gain and pattern. He had worked out two geometries that might be suitable for my use. One was a 33-element, 10.6-wavelength (24-ft) model, and the other was a 17-element, 4.5-wavelength (10-ft) version.

I modeled two different approaches on the computer. The first cut the 33-element Yagi to a 22-element version; the second extended the 17-element model to 21 elements. Both the 21- and 22-element models (which use different element spacings) would be about 14 ft long. This is the maximum boom length I felt safe with, considering that my antenna criteria called for a lightweight, low-windload and no-boom-support design.

I built examples of Tom's 33- and 17-element Yagis, but found they performed considerably worse than expected. Careful pattern measurements and further computer analysis indicated that the antennas were tuned too low in frequency. Revised versions of the 33-element Yagi (with shorter elements) gave measured performance near what the computer predicted. This indicated that the W1EJ designs were worth pursuing. It also showed how a computer-created design must go through post-computation measurement and adjustment to verify its performance.

Examination of adjusted computer models of the 21- and 22-element Yagis showed that, as with any engineering design problem, there were trade-offs between both designs—and no clear-cut winner. The 21-element model could be computer tweaked for more gain (15.9 dBd theoretical versus 15.8 dBd for the 22-element Yagi). The pattern on the 22-element design was easier to control, and it had a significantly smaller rear lobe. When test antennas were constructed, the 22-element version won because of its better pattern (important for EME work). Note that the 21- and 22-element designs are not the ultimate in gain, as their spacings were not specifically optimized for a 14-ft boom length. Optimization of spacings for such a Yagi of this specific length could take several months and produce no more than an additional tenth of a decibel in forward gain!

Tom's computer design work provided antenna dimensions given in tenths of millimeters. I spent several weeks adjusting the Yagi's geometry on the computer to create an easy-to-build version with dimensions given in US customary units that would retain the theoretical performance of the computer model in real life. In addition, I worked on tuning the Yagi's center frequency to the desired range. It took only two tries to build a real Yagi with acceptable performance from the computer model.

The finished 22-element antenna was presented at the New England and Central States VHF conferences in 1986. On my home antenna range, I measured an antenna gain of 15.7 dBd—0.8 dB better than an RIW 19. The front-to-back ratio (F/B) measured 20 dB (5 dB better than the RIW). At the New England conference, the measured gain was 0.6 dB better than the RIW; at the Central States conference, it measured 1.0 dB better. Overall pattern measurements showed the 22-element antenna to have a better pattern than the RIW 19.

Computer modeling calculates the 22-element antenna gain to be 0.9 dB more than the RIW 19 with the pattern improvements confirmed on the test range. When you use a computer program to optimize a design, you can never be sure that the design will work as expected. This is because all models have some errors caused by calculation assumptions, algorithm errors or just plain "multiple-calculation build-up" errors. If a design is optimized with an even slightly erroneous calculation, the resultant dimensions will incorporate those errors.

Further Optimization

I still wanted to try for a better pattern and more gain before building the new EME array. Another two months were spent further optimizing the design and reworking the dimensions for metric units.

I felt that metric units were appropriate if this was to be an antenna design of the 1980s and beyond. The final design has the same gain at 432 MHz (15.7 dBd) and an improved pattern. The peak gain of this Yagi is 1 MHz higher than the previous version (437 MHz). This was done to improve the pattern, assure excellent operation in large arrays and retain that performance in wet weather.

In its final form, no element length or spacing dimensions are the same. This seems to be characteristic of Yagis with maximum all-around performance. By maximum all-around performance, I mean a combination of a very clean pattern, excellent gain bandwidth and high gain for the boom length.

Resistive Losses

You may wonder where the missing 0.1 dB is between the calculated and measured Yagi gain. Such a small difference (3%) could easily be attributed to calculation error. After several years of building and measuring Yagis and comparing them to computer models, however, I have added correction factors to account for most of the difference.

Resistive losses account for most of the gain difference. Aluminum has an electrical resistance. Because current flows in all elements of a properly designed Yagi, losses will accumulate in the elements. DL6WU has shown that resistive losses are distributed fairly evenly throughout all elements. For maximum performance, the Yagi must be built from material with good conductive characteristics that will perform well in the weather.

Rainer Bertelsmeier, DJ9BV, has analyzed the K1FO 22-element Yagi with the sophisticated NEC3 program and calculated its resistive losses to be about 0.06 dB. Changing to copper elements would reduce these losses to 0.04 dB. My antenna is among the better designs in terms of resistive loss. Although lower losses are possible (0.04 dB is the lowest calculated by DJ9BV for a Yagi with similar gain using aluminum elements), lower-loss designs require greater boom length to achieve the same gain. There is no perfect solution. Part of the design problem is determining a tolerable resistive loss versus gain per boom length. The resistive-loss problem demonstrates another trap in computer analysis: It's possible to come up with a great theoretical design that may be a poor real-world performer because of resistive losses—this has occurred!

The other 0.04 dB difference between calculated and measured gain is caused by losses in the UT-141 balun. The use of an air-dielectric quarter-wave sleeve balun could reduce these losses to about 0.02 dB.

As a practical matter, of course, element and balun losses are not detectable in an antenna system used for terrestrial work. Even in an EME array, it will require the best of receiving systems to detect any improvement made from the reduction of these losses.

Pattern Measurements

Both the calculated and measured results demonstrate the value of the time spent in cleaning up the pattern. Figure 1 shows the calculated E- and H-plane patterns, and Figure 2 shows the measured E-plane pattern. The front-to-back ratio is 22 dB, and the first E-plane sidelobe is down about 17.5 dB.

Gain Bandwidth

Figure 3, a plot of calculated gain versus frequency, demonstrates the extremely wide gain bandwidth of the K1FO 22-ele-

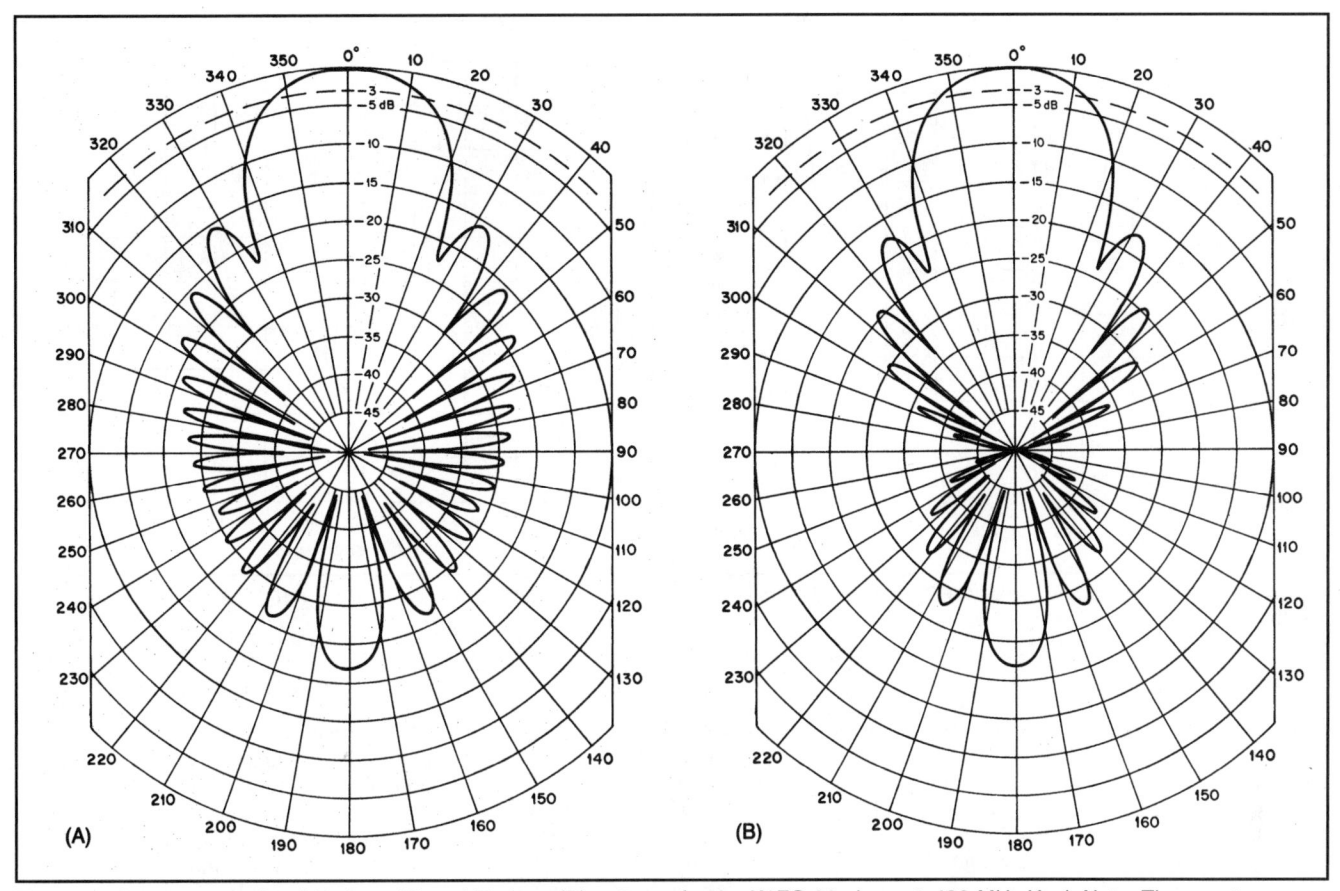

Figure 1—Computer-predicted H-plane (A) and E-plane (B) patterns for the K1FO 22-element, 432-MHz Yagi. Note: These antenna patterns are drawn on a linear dB grid, rather than the standard ARRL log-periodic grid. The linear dB grid shows sidelobes in greater detail and allows close comparison of sidelobes among different patterns. Sidelobe performance is important when stacking antennas in arrays for EME work.

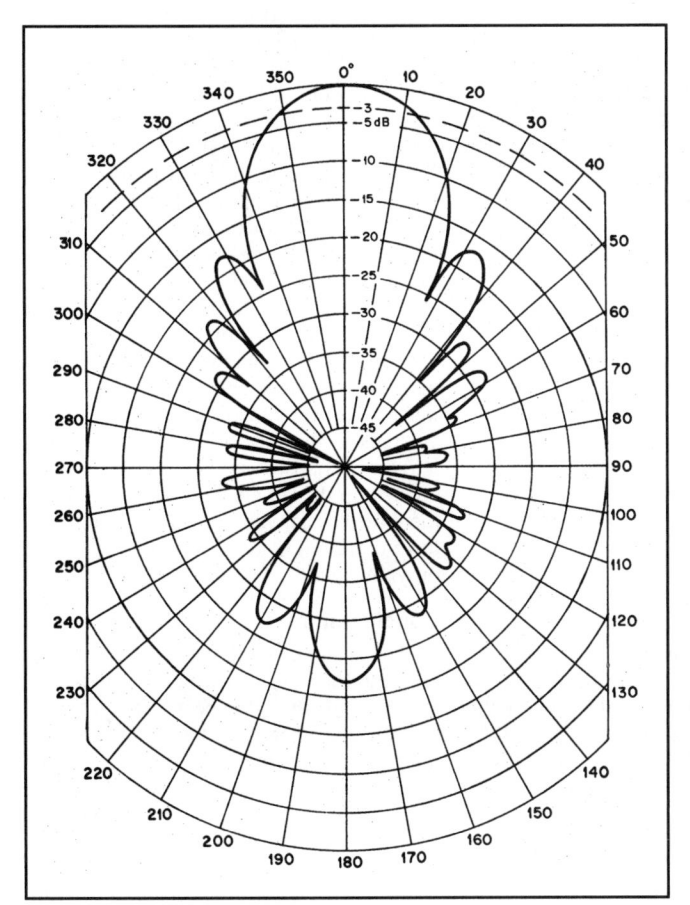

Figure 2—Measured E-plane pattern for the K1FO 22-element Yagi. Note: This antenna pattern is drawn on a linear dB grid, identical to the grids in Fig 1, rather than the standard ARRL log-periodic grid.

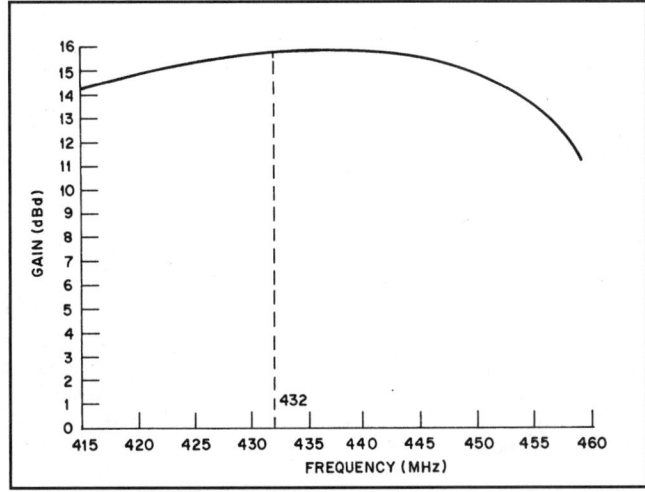

Figure 3—Gain versus frequency for the K1FO 22-element Yagi. Note that the 1-dB-gain bandwidth is 31 MHz and that the gain peak occurs at 437 MHz.

432 MHz 3-3

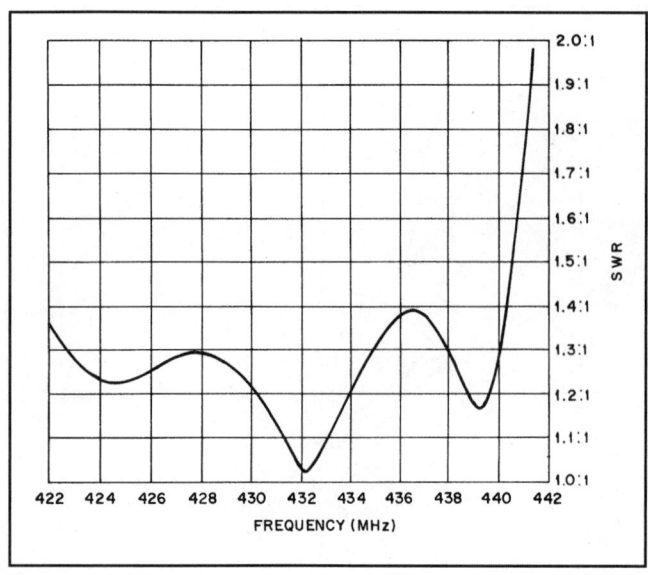

Figure 4—SWR performance of the K1FO 22-element Yagi in dry weather.

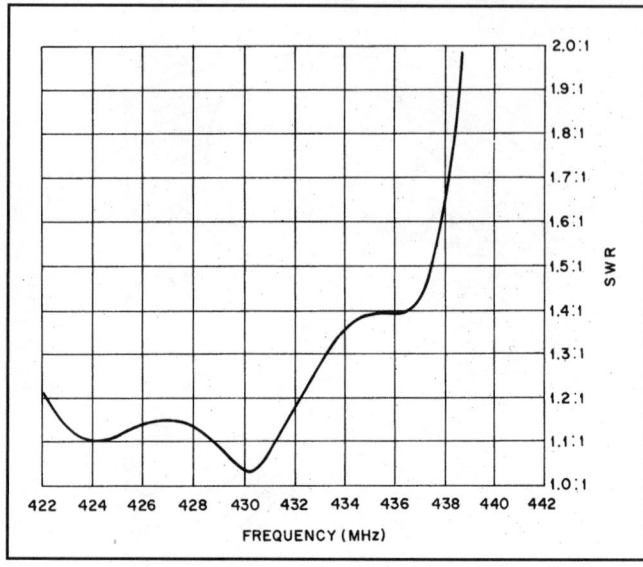

Figure 5—SWR performance of the K1FO 22-element Yagi in wet weather.

ment Yagi. Swept gain measurements of the Yagi using a network analyzer have confirmed the calculated gain bandwidth and center-frequency tuning. With an absolute gain peak at 437 MHz, the gain is less than 1.0 dB down between 420 and 450 MHz. The persistent myth that Yagis have very narrow bandwidths should be discredited forever.

Gain bandwidth (a measure of forward gain versus frequency) should not be confused with SWR bandwidth (SWR versus frequency). SWR bandwidth is a measure of the feed-point impedance and is not necessarily indicative of gain or pattern performance.

A wide gain bandwidth is important, even if operation will be on only a narrow band of frequencies. This is true for the following reasons:

1) *Construction tolerances.* The wider the bandwidth, the less critical the tolerances when building the Yagi. This makes it easier to retain excellent performance when duplicating antennas.

2) *Minimum shift of phase center.* A Yagi with smooth gain roll-off characteristics usually has less of a phase difference at the driven element as frequency changes. This is important in large arrays where the center Yagis will be operating at different points on their frequency response (this is caused by unequal mutual-impedance effects). Large phase shifts are characteristic of Yagis with many element spacings and lengths that are the same. It is also one of the major reasons why early amateur long Yagi designs were poor performers when used in large arrays.

3) *Array center frequency.* The value of wide gain bandwidth is related also to mutual-impedance effects. At 432 MHz, most of us are using arrays of Yagis (two or more antennas). Mutual-impedance effects tend to lower the center frequency of an array of Yagis relative to the center frequency of an individual Yagi. I have measured the drop in center frequency for an array of four RIW 19s to be about 400 kHz. Based on this experience, an array of 16 RIW Yagis might exhibit a center frequency drop of more than 1 MHz. An array made from wide-gain-bandwidth Yagis is a better choice than an array made from Yagis that exhibit a sharp gain drop on the high side of the peak gain frequency.

If the gain of a Yagi drops off rapidly just above the desired frequency of operation, lowering the array center frequency causes the stacking gain to be substantially lower than the theoretical 3.0 dB for doubling the array size. In addition, the pattern deteriorates rapidly above the maximum gain frequency for most Yagis (and especially for narrow bandwidth designs). For EME operation at 432 MHz, such poor Yagi pattern characteristics also create poor array patterns. This results in inferior receive performance because of unwanted earth-noise pickup.

SWR Bandwidth

A lot of time was spent designing a driven element that would have excellent dry weather and good wet weather SWR. I decided on a T match and optimized it at 432 MHz using a Hewlett-Packard 8753A network analyzer. A sweep of SWR versus frequency is shown in Figure 4. In dry weather, the SWR measured less than 1.10:1 from 431.2 MHz to 433.1 MHz.

The good SWR bandwidth results from the wide gain bandwidth of the Yagi and from tuning the director string above the center operating frequency. For this Yagi design (as well as for other designs that were built and tested), the driven element impedance changes less with frequency on the low-frequency side of the gain peak than at or above the gain peak.

I used a garden hose to wet the Yagi for simulated heavy rain conditions. A plot of SWR versus frequency under these wet conditions (Figure 5) demonstrates how the match center frequency shifts when the Yagi is wet. The network analyzer showed that when wet, the driven element impedance becomes more inductively reactive. The SWR at 432 MHz is still an excellent 1.18:1 when the Yagi is wet. My present array of 12 22-element Yagis measures (in the shack) well under 1.2:1 when dry, and about 1.3:1 in heavy rain. Icing is a different story. As with all other 432-MHz Yagis I have tested, performance is seriously degraded under icing conditions.

I have been asked if a quad-loop driven element could be used with this Yagi. I do not favor the use of a quad-loop driven element on a long Yagi. The only advantage of a quad loop is a slightly greater driven-element match bandwidth. All the old myths about lower noise pickup, higher gain and better pattern are exactly that: old myths. The quad-loop driven element also adds weight and windload area to the antenna without improving its gain. (I modeled this design on the computer with a quad-loop driven element and found that to make it work at all would require extensive changes to the first three or four director lengths and positions. If you want to try a quad-loop driven element or a driven-element feed system other than a T match, you have quite a job ahead of you, and you're on your own.)

Conclusions

The excellent pattern and gain of the 22-element Yagi are confirmed by the stacking spacings that give best array gain versus noise temperature. Optimum stacking distances are 65 inches in the E-plane and 62 inches in the H-plane. At those distances, stacking gain is almost 2.9 dB in both planes. (Phasing-line losses and gain loss caused by mechanical errors such as frame sag and misalignment are not included). The beamwidth of the Yagi at 432 MHz is 23 degrees in the E plane and 24 degrees in the H plane.

In examining the stacking characteristics of several popular 432-MHz commercial Yagis, I found it to be impossible to obtain more than 2.5 dB stacking gain (excluding phasing-line losses).[3] This phenomenon was not a function of the physics of stacking Yagis. It was caused by the design limitations of the Yagis under test. By comparison, calculated and measured stacking gains of approximately 2.9 dB (in both the E and H planes) were obtained with the K1FO 22-element design. This figure excludes phasing-line losses.

At 15.7 dBd gain for a 6.1-wavelength boom, the gain of the K1FO 22-element Yagi approaches the theoretical maximum for its boom length. It may seem that I have done a lot of work perfecting the design—and that a gain variation of a few tenths of a dB one way or the other isn't worth worrying about for a single 14-foot long Yagi. For an EME array of 8, 12 or 16 Yagis, however, the array gain versus wind load problem. makes the effort required to tweak the antenna very worthwhile!

Notes
[1]Powlishen, S., "High-Performance Yagis for 432 MHz," *Ham Radio*, Jul 87, pp 8-31.
[2]The first E- and H-plane sidelobes are actually a single cone-shaped lobe which surrounds the main lobe when the pattern is viewed in three dimensions.
[3]Powlishen, S., "Stacking Yagis is a Science," *Ham Radio*, May 85, pp 18-33.

By Steve Powlishen, K1FO

From *QST*, January 1988

An Optimum Design for 432-MHz Yagis

Part 2: Here is practical construction information for a high-performance antenna that you can build for that *big* signal on 432 MHz.

Last month, I described the development of the K1FO 22-element Yagi design. This month, I will give complete construction details and show how to scale the dimensions for other boom lengths.

Building the K1FO 22-Element 70-cm Yagi

I highly recommended that you follow exactly the construction information given here. If you use boom and element material of the sizes specified, you can build a top performing Yagi—provided you pay close attention to exactly duplicating the dimensions. Some builders will want to build the Yagi with whatever material is on hand. In addition to the specific information on my mechanical design, I have included some general guidelines for those who are willing to accept the problems faced when construction materials are changed. Please keep in mind that I cannot entertain requests to verify the dimensions, or assist in the construction or debugging, of Yagis that do not exactly follow the dimensions and construction materials recommended in this article. Translated, the last sentence says, "If you change anything, you're on your own!"

Boom Material

Figure 6 shows the boom layout. Round aluminum tubing of 7/8-inch and 1-inch outer diameter (OD) was chosen carefully to provide the highest possible boom strength, while maintaining low wind load and light weight. Round tubing has some disadvantages compared to square tubing: It is more costly to purchase and more difficult to drill accurately. Round tubing does have advantages, however: You can use it to make telescoping boom sections (allowing easy disassembly of the Yagi), and it offers lower wind load and lighter weight than square tubing. The boom could be made out of 3/4- or 7/8-inch-square tubing. The element lengths presented later do not have to be adjusted for square tubing of this size (provided that the same element-mounting method is used).

If you use tubing of a different diameter for the boom, you will need to apply a boom-correction factor. Lengthen each element by 1 mm for each 1/8 inch increase in boom diameter; shorten each element by 1 mm for each 1/8 inch decrease in boom diameter.[4] For example, if the entire Yagi is made of 1 1/4-inch tubing, the reflector, driven element and directors D1-D9 and D14-D20 should be 3 mm longer, while directors D10-D14 should be 2 mm longer. I don't recommend use of a boom larger than 1 1/2 inches OD—such sizes make com- pensation for boom effects difficult. Do not use square tubing smaller than 3/4 inch or round tubing smaller than 7/8-inch and 1-inch OD—these materials are not strong enough.

The boom materials shown in Figure 6 have been chosen for good strength versus weight. I recommend 0.049-inch-wall tubing for the 7/8-inch boom sections. Tubing with a 0.058-inch wall is suitable, but it is slightly heavier. The 1-inch center section should be 0.058-inch-wall tubing so that the end pieces telescope properly, with a minimum of slack. If a single mast clamp, mounted through the boom, is used, put a short piece of 7/8-inch tubing inside the center section where the mast clamp attaches. This doubles the wall thickness

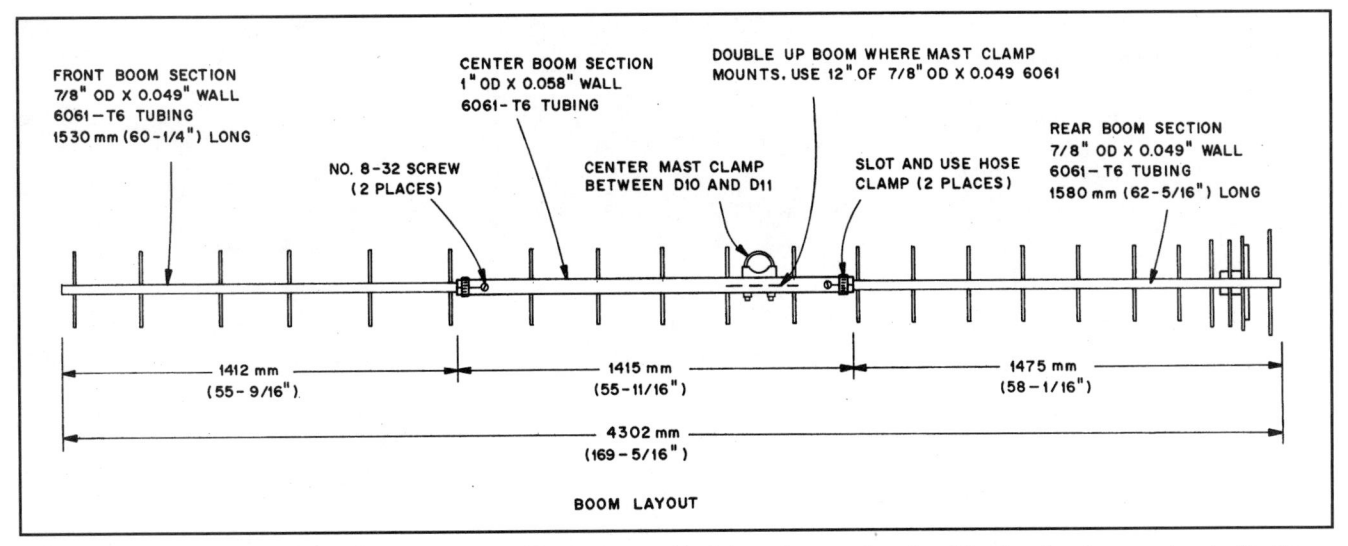

Figure 6—Boom-construction information for the K1FO 22-element Yagi. Lengths are given in millimeters to allow precise duplication of the antenna. See text.

Figure 7—Element-mounting detail. Elements are mounted through the boom using plastic insulators. Push-nut retaining rings hold the element in place.

for extra strength. My array of 12 22-element Yagis has already survived 1/2-inch ice loading in combination with winds stronger than 40 + mph.

The boom-section lengths were chosen so the antenna can be broken down and taken on mountaintop trips or to antenna-gain measuring contests. Unfortunately, these lengths result in a lot of wasted aluminum. An alternative construction technique (provided that you do not live in an area where you get heavy icing) is to make the 1-inch center section 48 inches long and appropriately lengthen the 7/8-inch sections. If you do this, remember to apply the boom-correction factors to adjust the lengths of any elements that were mounted in the 1-inch section, but are now mounted in the 7/8-inch section.

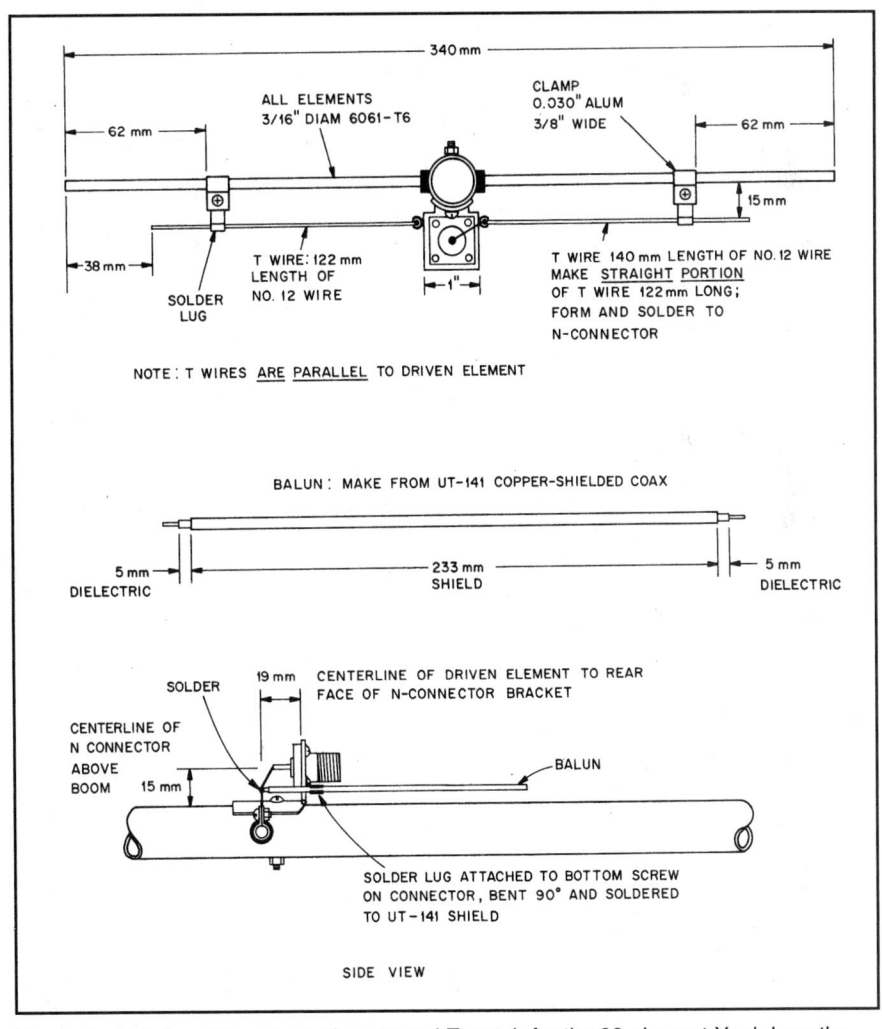

Figure 8—Details of the driven element and T match for the 22-element Yagi. Lengths are given in millimeters to allow precise duplication of the antenna. See text.

Elements are mounted through the boom with black delrin insulators held in place by push-nut retaining rings; see Figure 7. Nylon is an acceptable, but not as desirable, alternative—provided it is black to prevent it from deteriorating because of ultraviolet radiation. Teflon[R] insulators such as those described by George Chancy, W5JTL, are acceptable as well.[5] If the elements are mounted through the boom and not insulated, lengthen all elements by 5 mm (this applies for the 7/8-inch and

432 MHz 3-7

Table 1
Dimensions for the 22-Element 432-MHz Yagi

Element Number	Element Position (mm from rear of boom)	Element Length (mm)	Boom Diam (in)
REF	30	346	
DE	134	340	
D1	176	321	
D2	254	311	7/8
D3	362	305	
D4	496	301	
D5	652	297	
D6	828	295	
D7	1020	293	
D8	1226	291	
D9	1444	289	
D10	1672	288	
D11	1909	286	
D12	2152	285	1
D13	2403	284	
D14	2659	283	
D15	2920	281	
D16	3184	280	
D17	3452	279	
D18	3723	278	7/8
D19	3997	277	
D20	4272	276	

1-inch boom construction only). If the non-insulated mounting method is used, be sure that the elements have an excellent element-to-boom contact that will survive weathering effects.

To mount the elements above the boom in insulated blocks, follow the insulated-boom guidelines if the elements are centered at least 1/4 inch above the boom. If the elements are closer than 1/4 inch above the boom, they will require some lengthening. I am unable to give specific advice on lengthening the elements because I have not experimented with this construction method.

Use of nonconductive boom materials such as wood or fiberglass is not advised. In terms of strength versus weight and wind load, 6061-T6 and 6063-T6 aluminum tubing are without peer—at least for materials that amateurs can afford! Wood is a poor choice because of its short usable life and poor strength versus wind load and weight. Fiberglass is stronger than wood, but it is not as good as the high-strength aluminum alloys. In addition, fiberglass will deteriorate in sunlight unless it is protected from ultraviolet radiation. If you still insist on using a nonconductive boom material, shorten all elements in the 7/8-inch boom sections by 4 mm, and shorten those in the 1-inch boom sections by 5 mm.

Element Material

Element diameters other than 3/16 inch are not recommended. This size represents the best compromise between strength, weight, wind load and wet-weather de-tuning effects. For other element diameters, use the following guidelines—with caution!

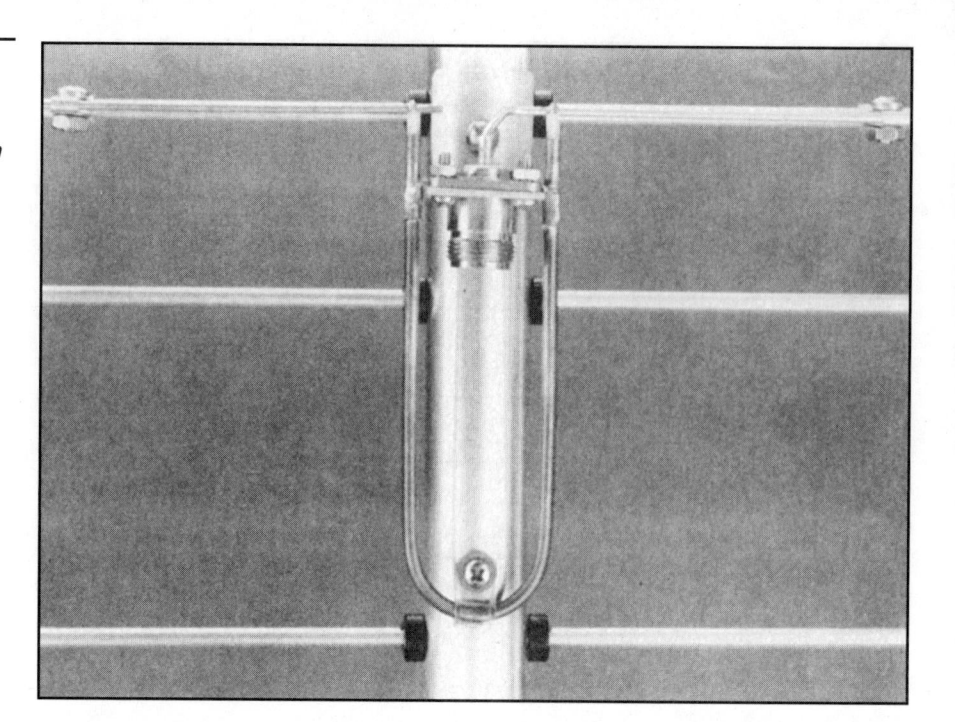

Figure 9—Various views of the driven element and T match.

Table 2
Design Information for K1FO Yagis of Different Lengths

Number of Elements	Boom Length (A)	Calculated Gain (dBd)	Element Number†	Base Element Length (mm)	Element-Length Correction (mm)	Last-Director Spacing (mm)
11	2.0	11.7	D9	289	−3	1444
12	2.4	12.2	D10	287	−3	1672
13	2.7	12.7	D11	285	−1	1909
14	3.1	13.1	D12	284	−2	2152
15	3.4	13.5	D13	283	−2	2403
16	3.8	13.9	D14	282	−2	2659
17	4.2	14.3	D15	281	−2	2920
18	4.6	14.6	D16	280	−1	3184
19	4.9	14.9	D17	279	−1	3452
20	5.3	15.2	D18	278	0	3723
21	5.7	15.5	D19	277	0	3997
22	6.1	15.7	D20	276	0	4272
23	6.5	15.9	D21	275	0	4550
24	6.9	16.2	D22	275	+1	4828
25	7.3	16.4	D23	274	+1	5109
26	7.7	16.6	D24	274	+1	5390
27	8.1	16.7	D25	273	+1	5672
28	8.5	17.0	D26	273	+1	5955
29	8.9	17.2	D27	272	+2	6239
30	9.3	17.4	D28	272	+2	6524
31	9.7	17.5	D29	271	+2	6809
32	10.2	17.7	D30	271	+2	7094
33	10.6	17.9	D31	270	+2	7380
34	11.0	18.1	D32	270	+2	7666
35	11.4	18.2	D33	269	+2	7952
36	11.8	18.4	D34	269	+3	8239
37	12.2	18.6	D35	268	+3	8526
38	12.7	18.7	D36	268	+3	8813
39	13.1	18.8	D37	267	+3	9100
40	13.5	18.9	D38	267	+3	9389

†Base dimensions for the reflector, driven element and directors D1-D8 are the same as those given in Table 1.

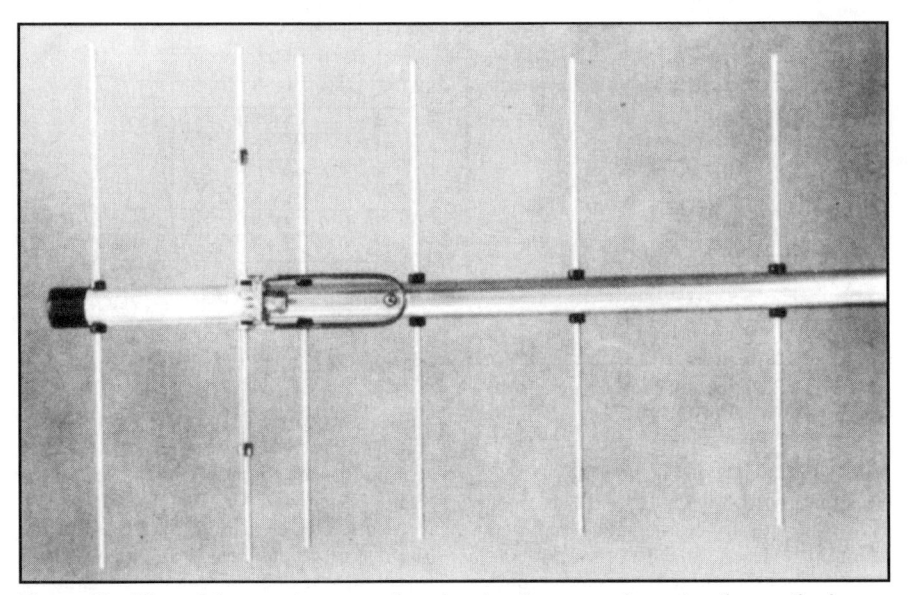

Figure 10—View of the rear boom section showing the general construction methods used.

- For 1/8-inch-diameter elements, lengthen all elements by 3 mm, but expect worse wet-weather performance and 0.1 dB less gain (caused by resistive losses). Skinny elements are also marginal from a mechanical standpoint. For these reasons, I recommend you stay away from 1/8-inch-diameter material.
- For 1/4-inch-diameter elements, shorten all elements by 3 mm. Resistive losses are theoretically 0.04 dB less with 1/4-inch-diameter elements, but the added weight and wind load may not make the larger size worthwhile.

Preparing the Boom and Elements

All element lengths and positions are given in metric dimensions, rather than US customary units. See Table 1. Metric dimensions are much easier to work with, especially for cutting and centering elements. If you plan a significant amount of antenna work, buy a good metric scale and tape measure. If you are stuck on using inches, try to keep to 1/32-inch tolerances when converting the given metric dimensions to US customary units.

Note that element positions are referenced from the reflector end of the boom. For example, the reflector position is not at 0, but at 30 mm. This makes it easy to mark the boom for drilling if you have a good tape measure longer than 4.3 m. Start by cutting all the boom sections to size. Next, slot the ends of the center section and use hose clamps to secure all three boom sections in place. You can then mark the element-drilling positions by putting the tape measure on the end of the boom and simply going down the boom and marking the element locations. Finally, scribe marks in the 7/8-inch boom section where they meet the center boom section, These marks make final boom assembly easier once the elements are in place.

Keep tolerances to ± 0.5 mm if possible. Because the antenna bandwidth is so great, gain is virtually unaffected by 1-mm measurement errors. The pattern, however, may deteriorate if you get too sloppy in your construction methods.

Elements are rough cut easily with a small hacksaw. Using a vise to hold the element stock makes the job much easier. Tin elements can then be filed to the exact dimension. For measuring element length, I lay a good machinist's scale flat on a workbench and butt it against a straight object (such as a metal bar). Then, I butt the element end against the straight object and mark the length with a sharp scribe. I have no trouble trimming elements to within 0.25 mm using this method.

To finish the elements, I use a file to put a 1-mm chamfer on each end. The chamfer is designed into the element lengths. I feel that it slightly improves wet-weather performance and makes it easier to start the push-nut element retainers.

Driven Element

The driven element and T match used on the K1FO Yagis is patterned after the driven element and match on the RIW Products 19-element Yagi. If you cannot figure out exactly how to build the driver element, find someone with an RIW 19 and take a look at it. Figure 8 shows dimensions for the driven element and T match. Figure 9 shows the driven element photographically from different views. Figure 10 shows the general construction of the rear boom section.

If you want to optimize the match for a frequency other than 432 MHz, adjusting the driven-element dimensions should not affect Yagi performance as long as the driven element does not get overly long—more than 343 mm. You can change the size and spacing of the T-match wires, but I do not recommend changing the balun

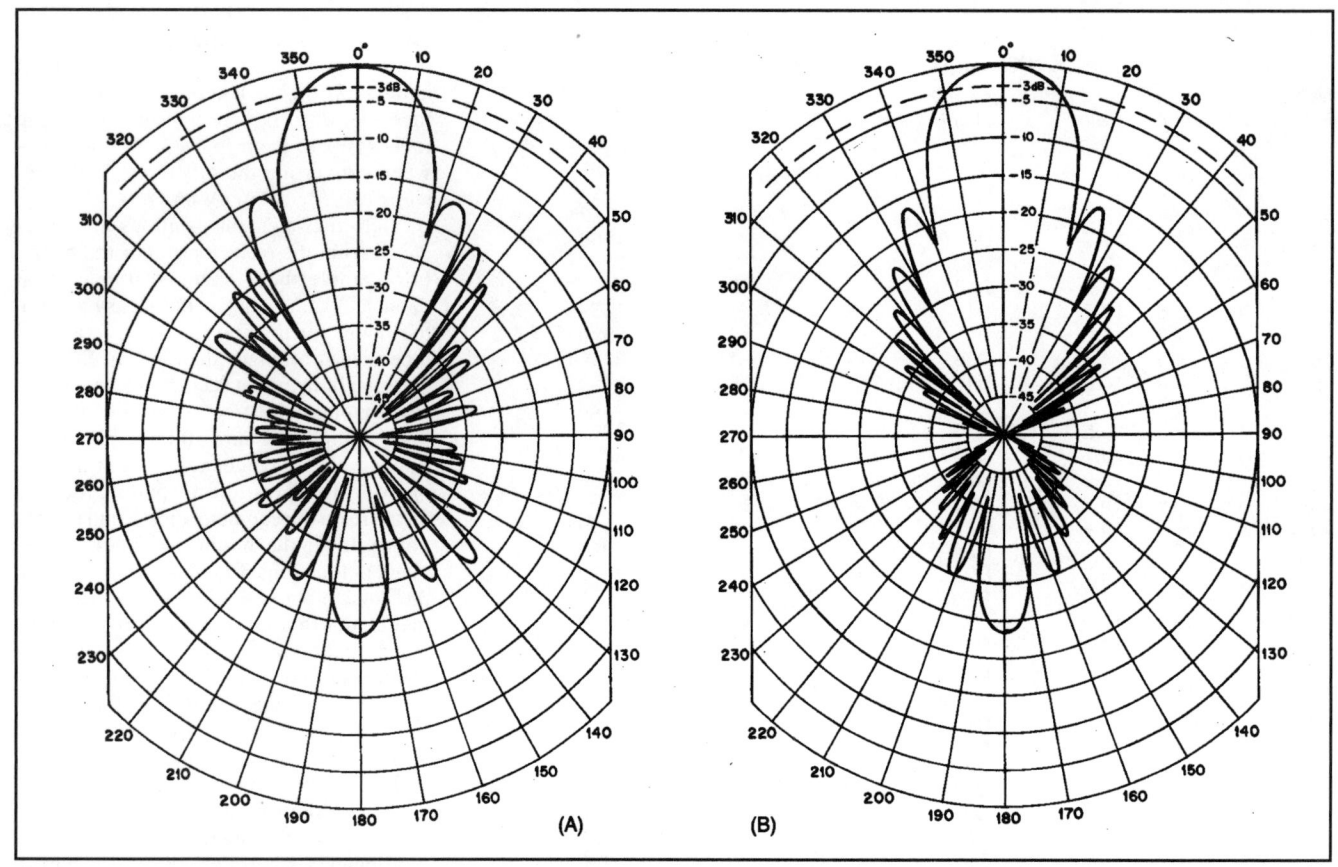

Figure 11—Measured (A) and predicted (B) E-plane patterns for the 33-element K1FO Yagi. Note: These antenna patterns are drawn on a linear dB grid, rather than the standard ARRL log-periodic grid. The linear dB grid shows sidelobes in greater detail and allows close comparison of sidelobes among different patterns. Sidelobe performance is important when stacking antennas in arrays for EME work.

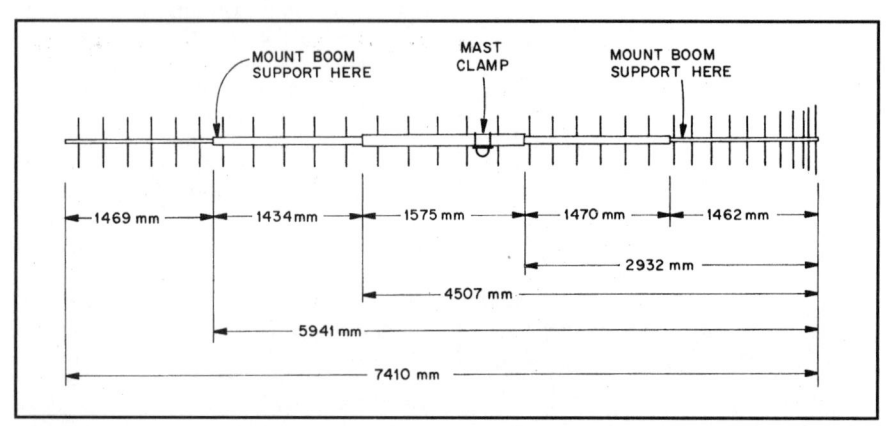

Figure 12—Boom-construction information for the K1FO 33-element Yagi. Lengths are given in millimeters to allow precise duplication of the antenna. See text.

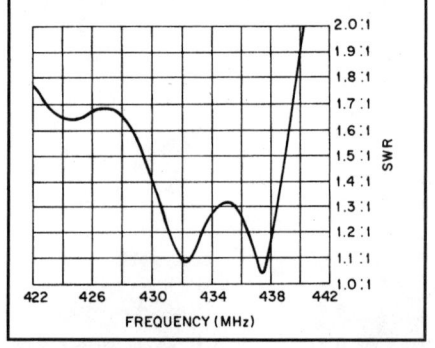

Figure 13—SWR performance of the K1FO 33-element Yagi in dry weather.

length. The balun length was chosen carefully to be an exact electrical half wavelength. Tests indicate that baluns of other lengths upset pattern balance. The first director position or length could also be adjusted slightly to improve the match, but don't change either dimension by more than 3 mm.

Making the Yagi for Other Boom Lengths

The variable-spacing geometry allows the K1FO 22-element Yagi to be scaled to other boom lengths. If the antenna is made significantly shorter or longer, adjustments to the element lengths are required. For versions with fewer than 11 elements (2.0-wavelength boom), gain will be considerably less than optimum. Gain improvement for these short Yagis requires optimization of the directors for the specific boom length. Gain is very good out to 40 elements (13.5 wavelength boom). Although the pattern remains excellent for all length Yagis, the first sidelobes do get somewhat stronger as the boom gets longer. To improve the pattern of the longer Yagis, you must optimize director lengths for the specific boom length. Driven-element tuning for an acceptable 50-ohm match is required for each version.

Table 2 summarizes performance and scaling information for Yagis based on the K1FO 22-element Yagi between 11 and 40 elements. The first three columns show the number of elements, the boom length in wavelengths, and the calculated gain for each version.

The *Base Element Length* column shows the base length, in millimeters, for directors D9 and above. (The base length for the reflector, driven element and directors D1-D8 are given in Table 1.) Note that two correction factors must be applied to these element lengths.

1) The *Element-Length Correction* column shows the amount to shorten or lengthen all elements, relative to base element lengths. For example, if you want to

3-10   Chapter 3

Table 3
Dimensions for the 33-Element 432-MHz Yagi

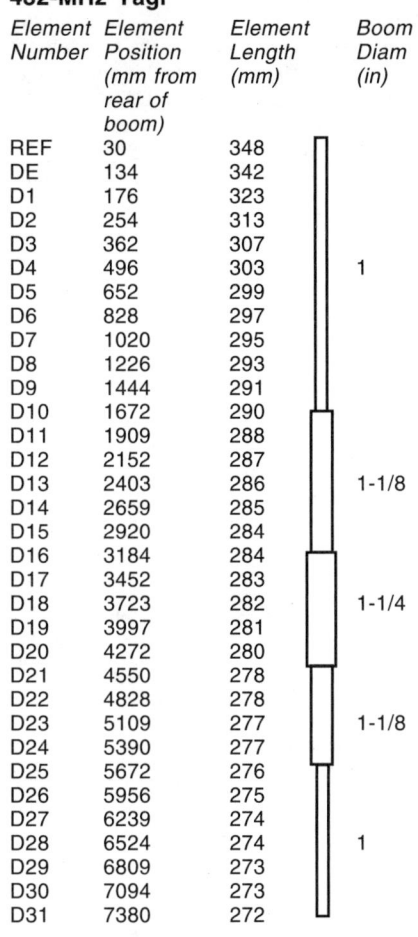

Element Number	Element Position (mm from rear of boom)	Element Length (mm)	Boom Diam (in)
REF	30	348	
DE	134	342	
D1	176	323	
D2	254	313	
D3	362	307	
D4	496	303	1
D5	652	299	
D6	828	297	
D7	1020	295	
D8	1226	293	
D9	1444	291	
D10	1672	290	
D11	1909	288	
D12	2152	287	
D13	2403	286	1-1/8
D14	2659	285	
D15	2920	284	
D16	3184	284	
D17	3452	283	
D18	3723	282	1-1/4
D19	3997	281	
D20	4272	280	
D21	4550	278	
D22	4828	278	
D23	5109	277	1-1/8
D24	5390	277	
D25	5672	276	
D26	5956	275	
D27	6239	274	
D28	6524	274	1
D29	6809	273	
D30	7094	273	
D31	7380	272	

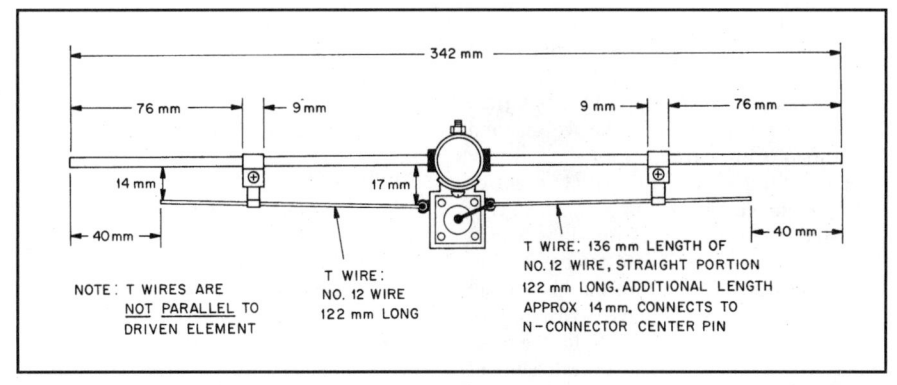

Figure 14—Details of the driven element and T match for the 33-element Yagi. See Fig 8 for additional information. Lengths are given in millimeters to allow precise duplication of the antenna. See text.

build a 12-element antenna, all elements must be cut 3 mm shorter than the base lengths given.

2) The base element lengths assume that all elements are mounted in a 7/8-inch-diameter boom. For strength, you will probably use a larger-diameter boom for longer Yagis. You *must* make *an additional* element-length adjustment if you mount the elements in a boom other than 7/8 inch diameter. Add 1 mm to the base element length for each 1/8-inch increase in boom diameter. Note that if you use a combination of boom diameters (for example, 7/8-inch tubing at the ends and 1-inch tubing in the center), the boom-correction factor is applied to the elements mounted in the 1-inch section only.

The *Last Director Spacing* column gives the spacing, in millimeters, from the reflector end of the boom to the position of the last director. Use this information for element-position information for antennas longer than 22 elements.

Here's an example of how to use Table 2. First, select the number of elements desired. Let's build a 7.3-wavelength, 25-element Yagi, with a calculated gain of 16.4 dBd. From the *Element-Length Correction* column, all elements must be lengthened 1 mm over the base dimensions.

Remember that the element lengths are for a 7/8-inch boom. From a structural standpoint, it may be desirable to make the 25-element Yagi boom from two 6-foot sections of 1-inch tubing telescoped into a 6-foot center section made from 1 1/8-inch tubing. The elements must be lengthened further for such a boom: All elements in the 1-inch boom sections must be lengthened another 1 mm, and the elements mounted in the 1 1/8-inch boom piece must be made another 2 mm longer. So, taking into account adding both correction factors, the elements in the 1-inch boom section must be a total of 2 mm longer than the base dimension, and the elements mounted in the 1 1/8 boom section must be a total of 3 mm longer.

A 33-Element Yagi

I built and tested a 33-element, 10.6-wavelength (24 ft, 3 in) Yagi from the information computed for Table 2. The theoretical gain of the Yagi is 17.9 dBd, and actual measured gain is closer to 17.7 dBd. Part of this difference appears to be explained by higher resistive losses in the 33-element Yagi, compared to the shorter antenna. The measured E-plane pattern of the 33-element Yagi (Figure 11A) is extremely clean and very close to the predicted pattern (Figure 11B).

Examination of the dimensions for this Yagi (Table 3) is also useful in determining how to adjust the Table 2 dimensions for other diameter booms and greater numbers of elements. Note that not all element lengths for the 33-element Yagi correspond exactly to the table. I adjusted some elements to optimize the pattern at this specific boom length and to achieve an excellent driven-element match.

Boom-construction details are shown in Figure 12. The boom for the 33-element Yagi starts out with 1-inch-OD × 0.049-inch-wall 6061 tubing. This telescopes into 1 1/8-inch-OD × 0.058-inch-wall tubing. A center section of 1 1/4-inch-OD × 0.058-inch-wall 6061 tubing completes the boom. Each of the five boom sections is approximately 5 feet long. This construction method increases the strength of the boom (to help eliminate sag and vibrations), and, as with the 22-element Yagi, makes the antenna easy to break apart for portable operation.

The 24-foot boom requires a support to minimize sag. Computer calculations indicate that the 3-4 inches of sag in the unsupported boom reduced antenna gain by 0.1 dB and caused significant distortion in the H-plane pattern. The support is made from a combination of 3/4-inch and 7/8-inch tubing. A 12-inch piece of 1 1/8-inch tubing, slipped inside the center boom section, strengthens the wall where the mast clamp mounts.

Like the 22-element antenna, the driven element on the 33-element Yagi was optimized with a sophisticated network analyzer. The longer Yagi also demonstrates excellent SWR bandwidth and an SWR at 432 MHz of close to 1.1:1 (see Figure 13). This is a good demonstration that a "gimmick" driven element is not needed to obtain a good match with a wide bandwidth on a long UHF Yagi. Note that the SWR is less than 1.33:1 for over 8 MHz. Wet-weather performance is also very good, with the center frequency shifting in a similar magnitude to the 22-element Yagi. Details of the long Yagi driven element are given in Figure 14.

Stacking distances for the 33-element Yagi have been calculated to be optimum at 85 inches E plane and 80 inches H plane. Stacking distances for antennas of other boom lengths can be interpolated from those calculated for the 22- and 33-element Yagis.

Variations on the K1FO 22-element design that are built with a significantly different number of elements may not work exactly as predicted. Although virtually any length Yagi should give excellent performance, some physical tweaking may be necessary to obtain maximum performance. Specifically, versions with boom lengths less than 4.6 wavelengths are generally 0.2 dB lower in gain than what is theoretically possible for such boom lengths. This is caused by the "universal" spacings that are used. The DE to D1 spacing is closer than needed for such short Yagis. In addition, some element length tweaks are needed to obtain the last few tenths of a decibel of gain on shorter versions.

Before we get too carried away with long Yagi designs, let's return to the original premise of this article. In a practical array, we must consider the weight and wind load of the array.

Let's compare two arrays: (1) eight of the 22-element, 14-foot-long Yagis; and (2) four of the 33-element, 24-foot-long Yagis. Measured gain is 15.7 dBd for the 22-element Yagi and 17.7 dBd for the 33-element Yagi. Phasing lines will have 0.3 dB loss for the eight-Yagi array and 0.2 dB for the four-Yagi array. This gives an overall gain of 24.0 dBd for the array of eight 22-element Yagis. The array of four longer Yagis has a total expected gain of 23.3 dBd—or 0.7 dB less than the eight-Yagi array.

Wind-load area of the 22-element Yagi is 0.8 square foot; the 33-element Yagi has an area of 2.8 square feet. The total array wind-load area must include the stacking frame, phasing lines, mounting plates and so on. When everything is included, the eight-Yagi array has a wind-load area of approximately 15 square feet. Wind-load area is 19 square feet for the four-Yagi array. Thus, the eight shorter Yagis have higher gain and less wind-load area! Even if the eight 22-element Yagis were arranged two wide and four high (so they would be better for use on terrestrial paths as well as for EME), the center of the array would not have to be mounted as far above the top guys as the four 33-element Yagis to allow for elevation movement. Note that if we wanted to build a four-Yagi array with gain equal to the eight smaller Yagis, we would have to use 37-element, 28-foot-long Yagis. The wind-load area of such an array would be almost 21 square feet.

Conclusion

I delayed publishing this information on the 22-element Yagi until I was sure that it performed as well as predicted. In October 1986, I replaced my 12 RIW 19 Yagis with 12 of the 22-element Yagis. I then spent two frustrating months coping with an array that never seemed to work the way it should. After a long string of problems (including water in two different phasing lines, a cracked shield on another line, and not one but two bad relays), the array is finally in full working order. The array uses the same phasing lines that were on the old array. Because these lines are a little short, the new array uses 64- × 60-inch spacing; the net array gain is 0.2 dB lower than the maximum possible for an optimally spaced array.

Sun noise is 15.0 dB during quiet sun periods, a solid 1.5 dB higher than with the old array. Earth noise (a measure of pattern quality independent of gain) is 5.0 dB, more than 0.5 dB better than the old array. Milky Way noise (the noise measured between cold sky and the center of our galaxy) is 5.3 dB.[6] Other celestial measurements are 3.0 dB on Cygnus, 2.9 dB on Cassiopeia and 1.2 dB on Taurus. These readings give an approximate total system temperature of 81 kelvins (K). Subtracting receiver noise (25 K) and phasing line noise (26 K), the total antenna noise is 29 K—a truly outstanding figure. Calculations by Rainer Bertelsmeier, DJ9BV, indicate an even lower array noise for the 22-element Yagi. More information on this subject can be found in an article by DJ9BV in the fourth 1987 issue of the West German VHF/UHF magazine, *DUBUS*.

The 22-element, 6.1-wavelength (14-ft) Yagi combines lightweight, low wind load, excellent gain for its size, a clean pattern and a wide gain bandwidth in one package. In addition, its geometry is adaptable to virtually any boom length. If you don't have the facilities to drill booms or cannot locate the parts required, Tom Rutland, K3IPW, makes available Yagi kits and components for the 22-element, 33-element and variations on this design.[7] My thanks to Tom Kirby, W1EJ, for his hard work in determining the basic design geometry for these Yagis.

Notes

[4] The actual correction in element length for each $1/8$ inch of boom diameter change is 0.8 mm ($1/32$ inch). If a boom size is used that is significantly larger than $7/8$ inch, the exact correction (0.8 mm) should be used.

[5] G. Chaney, W5JTL, "PTFE VHF Antenna Insulators," *Ham Radio*, Oct 85, pp 98-101.

[6] Sagittarius is the constellation used for array aiming. Because of the low elevation and nonpoint-source nature of the Milky Way, comparison of noise readings between different stations at different locations is not always meaningful.

[7] Rutland Arrays, 1703 Warren St. New Cumberland, PA 17070.

A 432-MHz Yagi for $9

This 22-element Yagi was designed with OSCAR Phase III in mind. Use of readily available materials makes it an attractive project.

Access to inexpensive but efficient equipment is a must if new technology is to remain within reach of the average amateur. Without returning to the oatmeal box and "cat's whisker," there are ways to keep our hobby an affordable one. This antenna is a step in that direction.

If all new material must be purchased, the cost of building this 22-element antenna should be about $9. (A complete list of necessary materials is presented in Table 1.) It is simple to build, weighs only 21 ounces,[1] and rolls into a compact package for storage or transport. A simple and inexpensive means may be employed to change polarization. The estimated antenna gain is 14.5 dB.

Construction

Refer to Figures 1 and 2 and the photographs. Antenna element lengths and spacings given in Table 2 were selected from information in the 1982 Handbook.[2] Gamma matching is used for easy adjustment. The gamma-matching capacitor I used is a tubular ceramic type that is easy to weatherproof with a silicone sealant after matching adjustments are completed. Using copper wire for the driven element and gamma rod allows the use of solder during assembly.

Table 1
Materials List

- 19 ft — no. 9 aluminum wire (clothesline wire).
- 13 in. — no. 8 copper wire.
- 6 in. — no. 14 copper wire.
- 28 ft — insulating spacers (insulation stripped from no. 14 wire or equivalent).
- 21 in. — 1/2.in. schedule 40 Pvc pipe.
- 1 — 1/2-in. schedule 40 Pvc T.
- 40 ft — 50-lb-test monofilament nylon fishing line.
- 2 — snap swivels.
- 1 — SO-239 chassis connector.
- 2 — plastic spreaders, 6 x 3/8 x 3/16 in.

Table 2
Antenna Element Lengths

Element	Length
Reflector	13-3/32 in.
Driven	12-23/32 in.
1st director	11-27/32 in.
2nd director	11-5/8 in.
3rd director	11-1/4 in.
4th director	11 in.
5th director	10-29/32 in.
6th director	10-13/16 in.
7th to 20th director	10-11/18 in.

Note: The reflector to driven element spacing is 5½ in. Spacing between all directors is 8⁷⁄₁₆ in.

Close-up view of the gamma matching section and driven element. Use of copper wire makes the assembly easy.

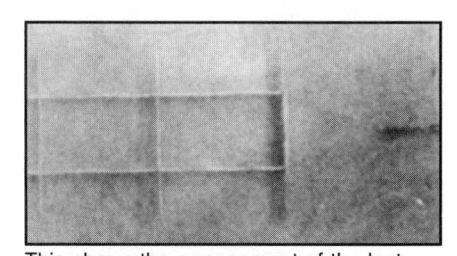

This shows the arrangement of the last three directors on the nylon fishing-line boom. The elements are separated by spacers fashioned from pieces of insulation removed from lengths of no. 14 wire. One of the plastic spreaders is next to the last director.

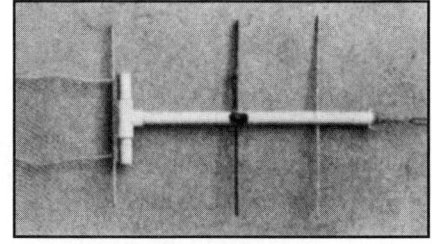

A short PVC boom supports the reflector and driven element. The PVC T at the left provides a means of attachment to the rest of the antenna.

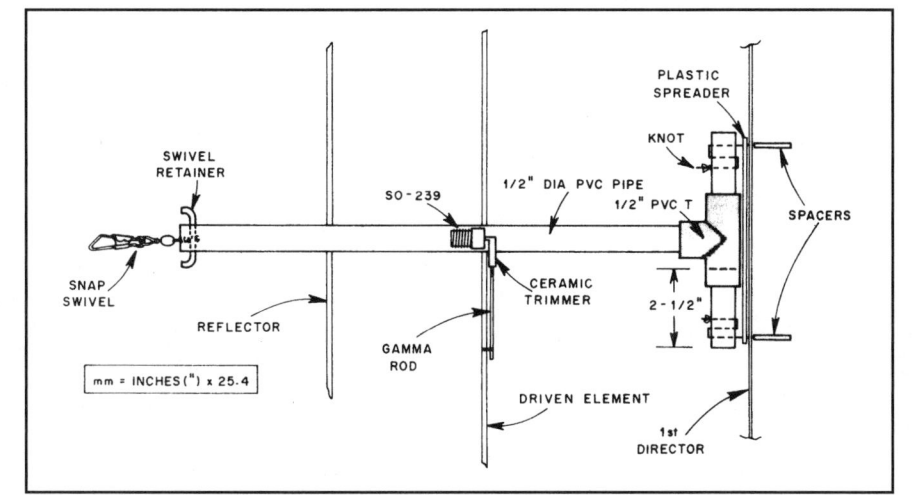

Figure 1—Mechanical assembly details of the PVC boom.

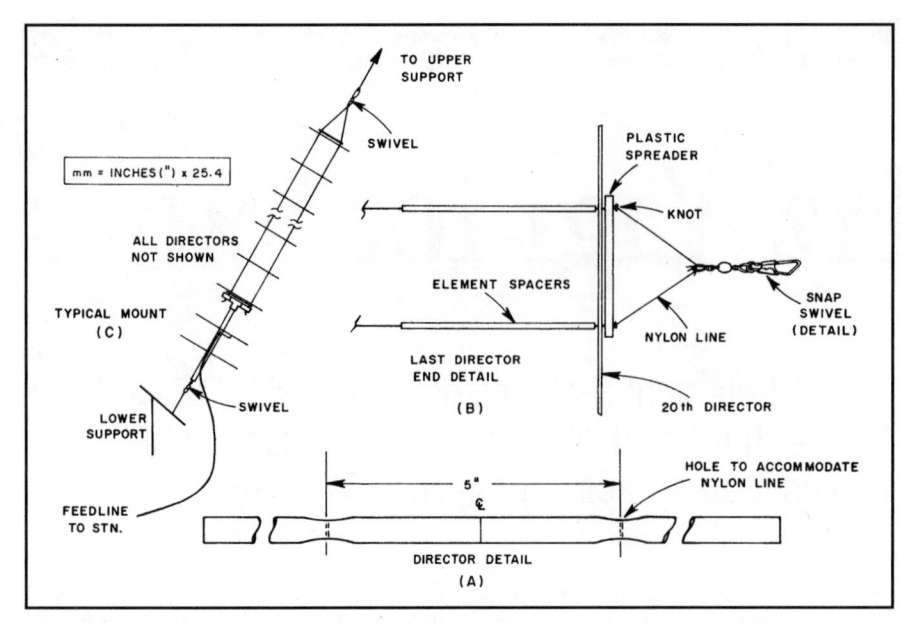

Figure 2—At A, the method used to fashion the director elements. A detail of the assembly at the 20th director is shown at B, while a typical method of installation is depicted at C.

The PVC pipe and T are assembled as shown in Fig. 1. This boom supports the copper wire driven element and single reflector, the latter of which is made of no. 9 aluminum wire. The assembly is fastened behind the first director as shown. An S turn of nylon line at each attachment point is used for tension adjustments. Snap swivels at the antenna ends permit it to be rotated easily to change polarization. The lower end of the antenna can be positioned to adjust antenna elevation and azimuth.

Results

All antenna testing was done using Mode J (2-meter uplink and 70-cm downlink) receive. In every case, the home-made antenna outperformed an 11-element Yagi with a preamplifier mounted on the boom. (Without the preamplifier, the 11-element Yagi produced signals that were barely discernible. Use of the preamplifier provided readable signals.) For both horizontal and vertical polarization of the antenna, the indicated beamwidths are 30°.

Now that you've got a simple and inexpensive way to build a gain antenna, why wait any longer? Let's hear you through OSCAR!

Notes
[1]g = oz × 28.35; mm = in × 25.4; kg = lb × 0.454.
[2]Element lengths were derived from Table 4, p. 21-5 and element spacings from Fig. 4, p. 21-4.

All directors are made from no. 9 aluminum wire and are slightly flattened at a point 2½ inches each side of center. Holes are drilled through the elements at these points to pass the 50-lb nylon fishing line that is used as a boom.

Cut the element spacers to provide the proper spacing as given in Table 2. Two scrap plastic spreaders are cut to a length of about 7 inches and drilled to pass the nylon line. One spreader is used at each end of the antenna. The insulated spacers and antenna elements are placed on the nylon boom as you would beads on a string.

Chapter 4
902 MHz

A 902-MHz Loop Yagi Antenna   Don Hilliard, WØPW   4-1

By Donald L. Hilliard, WØPW
From QST, November 1985

A 902-MHz Loop Yagi Antenna

Join in the fun on the newest Amateur Radio band with this fine performer that's also easy to build.

Several years ago, M. H. Walters, G3JVL, designed and published information on a loop-Yagi antenna for 1296 MHz.[1] Since that time, numerous modifications to his design have been published in the RSGB journal, *Radio Communication*.[2-4] These changes have mostly been minor ones. For some time, I have used one of the variants of G3JVL's design on 1296 and have been pleased with the results.

A few years ago, while studying various designs for 902-MHz antennas, I decided to scale the dimensions from the 1296-MHz loop Yagi and build one to check its performance. This article describes the results of my efforts and provides all of the information needed for you to build one for yourself.

Making the Boom

One of the first things to consider is the boom. I decided to make the antenna from readily available materials as much as possible. In most areas of this country, building-materials stores have various sizes of moderately priced aluminum tubing in 6- and 8-foot lengths. I decided to use two 6-foot lengths of 1-inch-OD tubing spliced to give me a 12-foot-long boom. The splice is made from a 2-foot length of 7/8-inch OD tubing that extends 1 foot inside each of the two 1-inch-OD pieces. This technique yields a strong boom that is adequate for this application. Although I have not tried it, I think a 12-foot length of 1-inch-square aluminum tubing would work as well and would make drilling the element-mounting holes much easier.

The antenna is mounted to the mast with a gusset plate. This plate mounts at the boom splice, and its mounting bolts also secure the splice (see Figure 1). Drill the gusset-plate mounting holes perpendicular to the element-mounting holes, assuming the antenna polarization is to be horizontal.

Once you have a 12-foot boom and mounting plate, you need some method of securing the boom while you drill the element-mounting holes. It's important to make the elements line up as closely as possible. Many gadgets that make this process

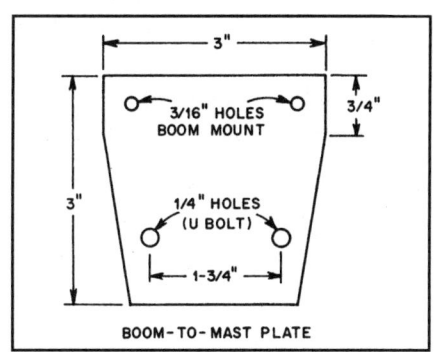

Figure 1—Boom-to-mast details.

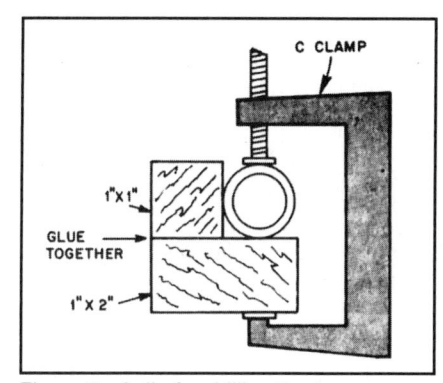

Figure 2—A Jig for drilling the boom can be made from some lumber and C clamps.

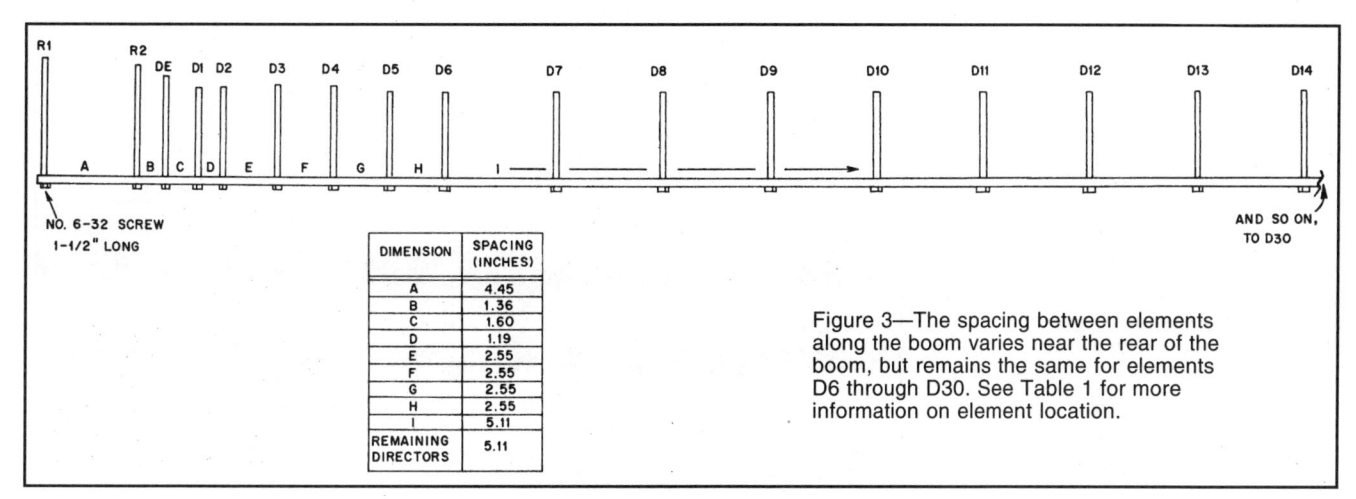

Figure 3—The spacing between elements along the boom varies near the rear of the boom, but remains the same for elements D6 through D30. See Table 1 for more information on element location.

DIMENSION	SPACING (INCHES)
A	4.45
B	1.36
C	1.60
D	1.19
E	2.55
F	2.55
G	2.55
H	2.55
I	5.11
REMAINING DIRECTORS	5.11

902 MHz 4-1

Table 1
Distance of Element-Mounting Holes from Reflector End of Boom

Element	Distance (Inches)	Element	Distance (Inches)
R1	0.50	D14	60.18
R2	4.95	D15	65.29
DE	6.31	D16	70.40
D1	7.91	D17	75.51
D2	9.10	D18	80.62
D3	11.65	D19	85.73
D4	14.20	D20	90.84
D5	16.75	D21	95.95
D6	19.30	D22	101.06
D7	24.41	D23	106.17
D8	29.52	D24	111.28
D9	34.63	D25	116.39
D10	39.74	D26	121.50
D11	44.85	D27	126.61
D12	49.96	D28	131.72
D13	55.07	D29	136.83
		D30	141.94

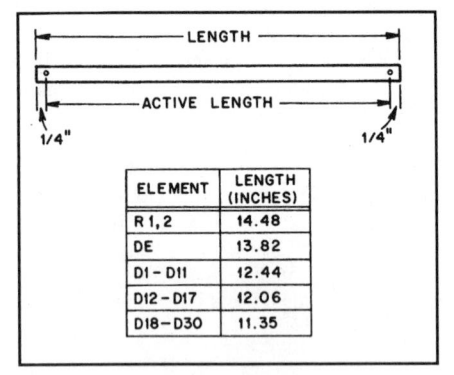

ELEMENT	LENGTH (INCHES)
R 1,2	14.48
DE	13.82
D1–D11	12.44
D12–D17	12.06
D18–D30	11.35

Figure 4—The active length of each element is 1/2 inch less than the length of the element strap. Lengths given here are the overall length of the element straps, not the active length.

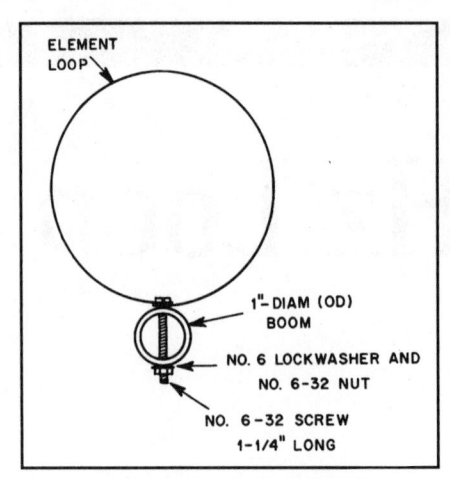

Figure 5—Element-to-boom mounting details.

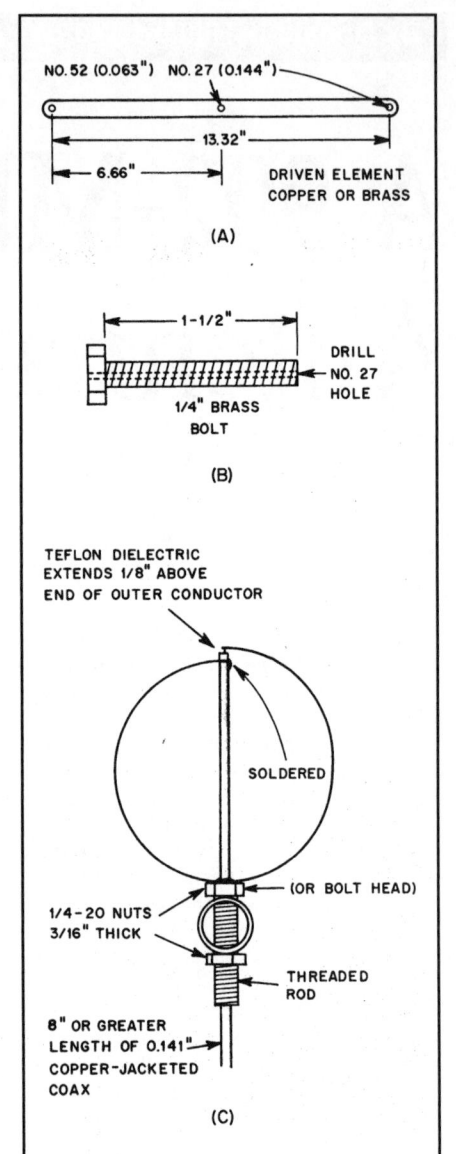

Figure 6—Driven-element details. The element strap is shown at A. At B is the brass mounting fixture. Part C details the driven element assembly.

easier than it might have been several years ago are available today. I use a jig constructed of two 12-foot lengths of pine lumber; one is a standard 1 × 2 and the other a standard 1 × 1. The boom is secured to the jig with three or four C clamps, and the holes are made with a drill press (see Figure 2).

Table 1 indicates the distances of the element-mounting holes from the reflector end of the boom. Fig. 3 shows the relationship of the various elements in the finished antenna. The element-mounting holes are made with a no. 28 drill to clear no. 6-32 mounting hardware. After you drill all of the holes, drill out the hole for the driven element assembly to 1/4 inch.

Making the Parasitic Elements

The reflectors and directors are cut from 1/32-inch (0.03125-inch) aluminum sheet and are 5/16 inch wide. Figure 4 indicates the lengths for the various element straps. They should be cut with a shear. Although the elements could be cut with tin snips, amateurs who have tried this have been disappointed in the results—badly distorted straps. It's much better to locate a friend with a shear, or have the job done at a machine shop.

Once you have the straps in hand, drill the mounting holes as detailed in Figure 4. The holes are made with a no. 28 drill, 1/4 inch in from each end of the element strap. After the holes are drilled, you must form each strap into a circle. This is done easily by wrapping the element around a round form. I used a 3-inch-diameter propane bottle.

Mount the loops to the boom with no. 6-32 machine screws, lock washers and nuts (see Figure 5). It's best to use stainless-steel or plated-brass hardware for everything. Although the initial cost is higher than for ordinary plated-steel hardware, stainless or brass hardware won't rust and need replacement after a few years. When you mount the elements, make sure they are perpendicular to the boom.

Making the Driven Element

Read this entire section carefully and study Figures 6 and 7 before you start assembly of the driven element. The driven element is cut from 1/32-inch (0.03125-inch) copper or brass sheet and is 5/16 inch wide. Drill three holes in the strap, as detailed in Figure 6A. Trim the ends to form a semicircle, leaving very little metal outside of the end holes. Form this strap into

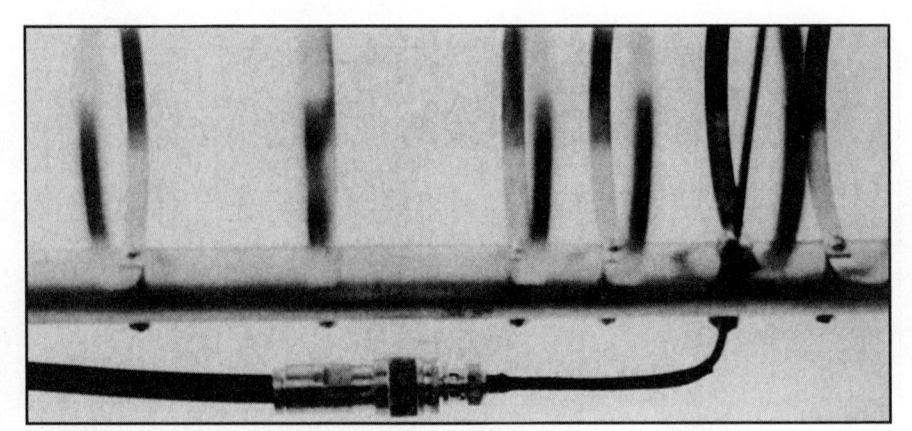

Fig. 7—A crimp-on BNC connector is soldered to the 0.141-inch Hardline that attaches to the driven element. An adapter is used to make the connection to the cable that runs down the boom to the main feed line.

(A) **(B)**

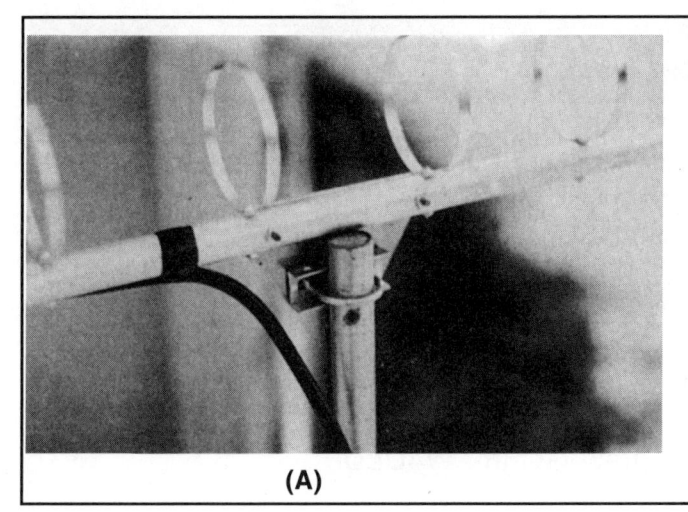

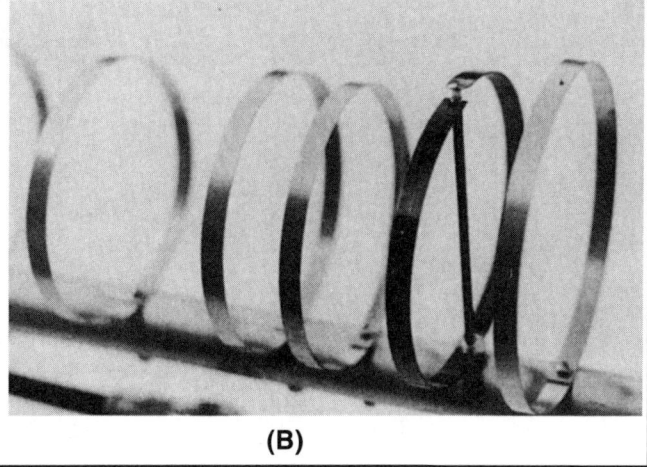

Details of the boom-to-mast plate and element mounting may be seen at A. A view of the driven-element assembly is shown at B.

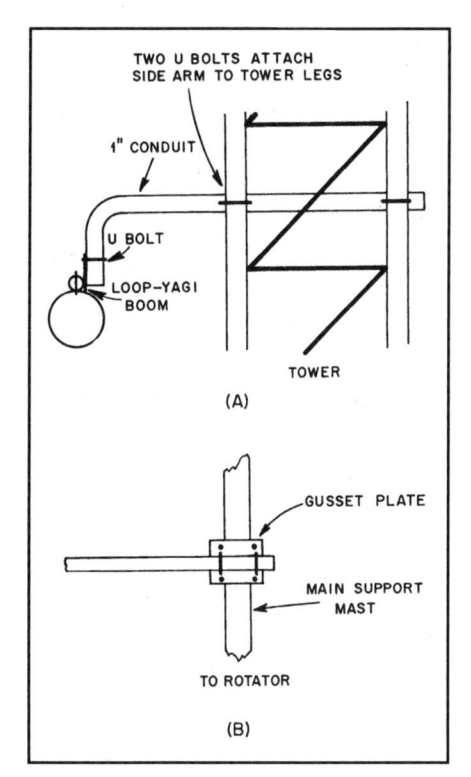

Figure 8—Suggestions for a fixed (A) or rotatable (B) mounting.

a loop similar to the other elements.

To mount the driven element to the boom, you need to make a brass mounting fixture, as shown in Figure 6B. The mounting fixture can be made from a 1½-inch brass bolt or from a piece of ¼-20 threaded brass rod. Bore a 0.144-inch (no. 27 drill) hole lengthwise through the center of the rod. A piece of 0.141-inch semirigid Hardline will mount through this hole and connect to the driven loop. Figure 6C shows how the driven element is assembled and mounted to the boom. The point at which the 0.141-inch Hardline passes through the copper loop and brass mounting fixture should be left unsoldered at this time to allow for matching adjustments when the antenna is completed.

The 0.141-inch semirigid Hardline, RG-402 or equivalent, is available in short lengths from varied sources. One source that sells this cable in foot lengths is J. Smith and Associates, Inc., 3540 N. Academy Blvd., Colorado Springs, CO 80907.

I use a jeweler's saw to prepare the cable ends. Only a few inches of this line are required, enough to get through the boom and mount a connector on it. At this frequency, you should use a minimum number of connectors or adapters; they frequently cause problems. Type-N cable connectors for RG-402 are available from various sources. One (Model no. 2707F) is made by Midwest Microwave, 3800 Packard Rd., Ann Arbor, MI 48104. Another source is Pasternack Enterprises, Coaxial Products Division, P.O. Box 26759, Irvine, CA 92713. A catalog that lists several suitable connectors for this application is available.

Or you may wish to make a connector the way I did, with a BNC connector and BNC-to-N adapter. I used a BNC crimp connector designed for use with RG-59 cable. The RG-402 outer conductor slips tightly inside the crimp body. The connector center pin may need to be drilled out to accept the RG-402 center conductor, or you can use the center pin from a BNC connector designed for RG-58 cable. The center conductor of RG-58 is the same size as that of RG-402, so an RG-58 center pin will work without modification. Solder the RG-402 copper jacket to the BNC crimp body and place a piece of heat-shrinkable tubing over the soldered joint to keep moisture and contaminants away. Fig. 7 shows the finished connection.

A low-loss RG-8 cable may be run down the boom and mast to the main feed line. For best results, your main feed line should be the lowest-loss 50-ohm cable obtainable. This cannot be overemphasized!

Tuning the Driven Element

Check the SWR of the antenna. You may realize some improvement by making the driven element more oblong. The antenna shown here exhibited an excellent match with no adjustment to the relative shape of the driven element. When you have obtained the desired match, solder the point where the RG-402 jacket passes through the loop and brass mounting fixture.

Mounting the Antenna

Now that the antenna is complete, it is ready to be mounted for use. You've probably noticed that the loops are not inherently strong. If you live in an area where there are no large birds, you may be able to mount it as shown in the photographs. I mount the single loop Yagi as shown in Figure 8. The inverted loop Yagi provides a less desirable perch for birds. Other deterrents have been tried, such as attaching toy snakes so that they move in the wind. A U-shaped (inverted) length of conduit will allow two of these antennas to be stacked horizontally for additional gain. The stacking distance should be 26 to 32 inches.

The loop Yagi may be used as a vertically polarized antenna. Simply rotate the boom on the gusset plate until the loops are horizontal. Be careful when you mount — any mounting member that is of the same polarization of the antenna must be kept a couple of wavelengths away from the antenna itself.

The gain of loop Yagis of this design has been measured at between 17 and 18 dBi. Stacking two of them will yield another 2.5 dB, approximately. Similar designs have used up to 16 stacked loop Yagis at 1296 MHz; these arrays have performed well enough to make EME contacts. It is an antenna whose performance is proven. I have the antenna shown in the photographs installed at my Missouri location (near Joplin). Look for me on 902-MHz CW!

Notes
[1] G. R. Jessop, ed., *VHF/UHF Manual*, 4th ed. (Hertfordshire: RSGB, 1983).
[2] D. Evans, "A Long Quad Yagi for 1296 MHz," *Radio Communication*, Jan. 1975, pp. 24-25.
[3] D. Evans, "The G3JVL Loop Yagi," *Radio Communication*, July 1976, p. 525.
[4] C. Suckling, "The G3JVL Loop Yagi Antenna," *Radio Communication*, Sept. 1978, pp. 782-783.

Chapter 5
Multiband

An LPDA for 2 Meters Plus	L. B. Cebik, W4RNL	5-1
A 146- and 445-MHz J-Pole Antenna	Andrew Griffith, W4ULD	5-6
An Easy, On-Glass Antenna With Multiband Capability	Robin Rumbolt, WA4TEM	5-10
A VHF/UHF 3-Band Mobile Antenna	J.L. Harris, WD4KGD	5-13
The DBJ-1: A VHF/UHF Dual-Band J-Pole	Edison Fong, WB6IQN	5-15
Another Way to Stack VHF/UHF Yagis	Brian Beezley, K6STI	5-18
An Easy Dual-Band VHF/UHF Antenna	Jim Reynante, KD6GLF	5-21

By L.B. Cebik, W4RNL
From *QST*, October 2001

An LPDA for 2 Meters Plus

Here's a high-performance 2-meter antenna with a nice bonus—it also covers 130-170 MHz, for your monitoring pleasure!

Building a log-periodic dipole array (LPDA) to cover all of 2 meters with good gain and a low 50-Ω SWR from 144 to 148 MHz requires about 6 elements and a 54-inch boom. The project would not make much sense (apart from satisfying raw curiosity), however, since that same number of elements on the same length boom can be arranged as a wideband Yagi with at least a dB more gain and an even lower SWR across the band.[1]

LPDAs find their niche wherever we need a wide operational passband with a relatively constant feed-point impedance. In the HF region, we typically build LPDAs for a 2:1 frequency range—for example, 14 to 28 MHz. Antennas with wider (10 to 30 MHz) and narrower (18 to 30 MHz) ranges are common. At VHF people have built wide-ranging LPDAs but most suffer from inadequate performance, except perhaps for general-utility purposes.

You can construct a fairly narrowband

The LPDA with elements oriented horizontally.

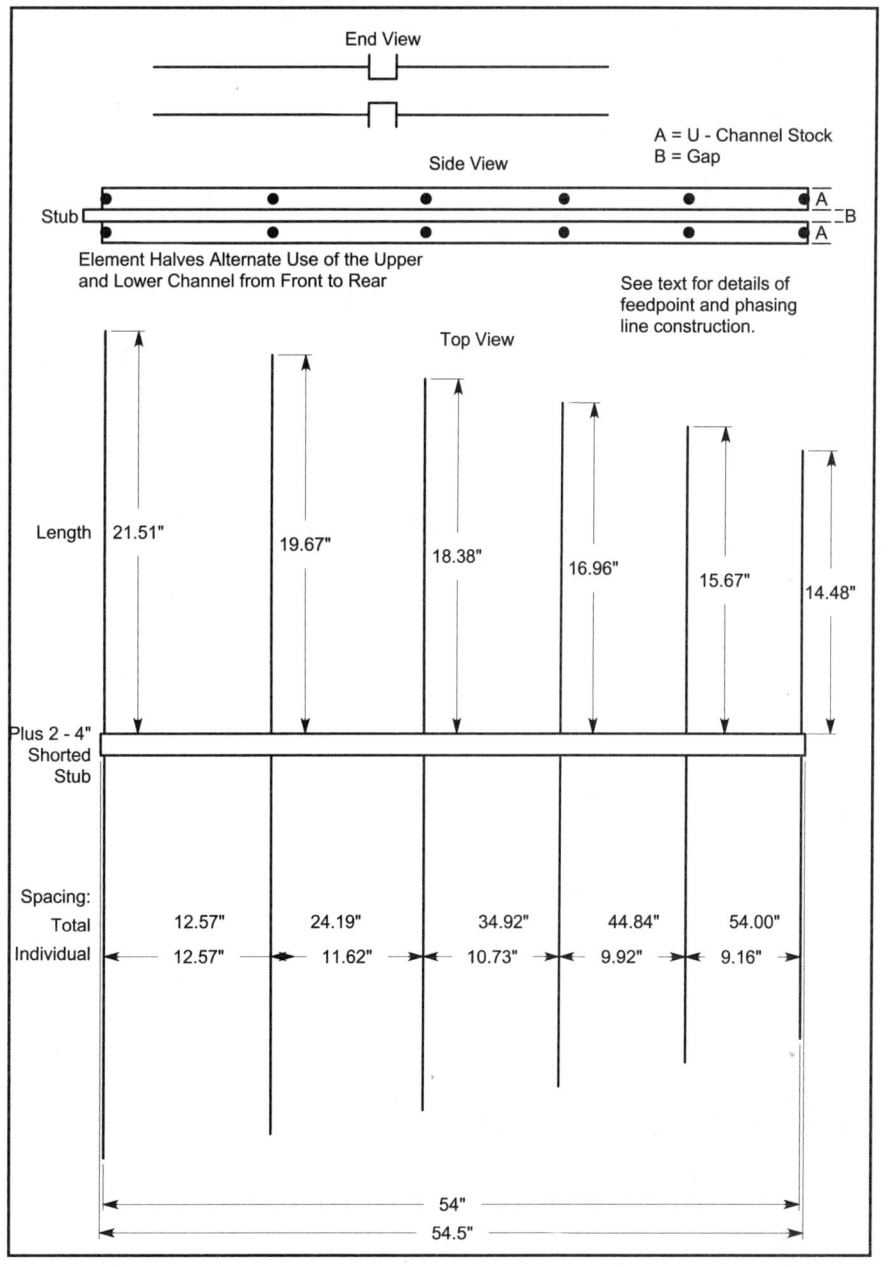

Figure 1—Outline sketch and dimensions of the 2-meter-plus LPDA.

LPDA centered on 2 meters and built to high performance standards. It will also offer something beyond the range of Yagi performance—the ability to monitor frequencies from 130 to 170 MHz with a 2:1 or better 50-Ω SWR. However, the lowest SWR values and the best performance (in terms of gain and front-to-back ratio) occur within the transmitting region, namely the 144-148 MHz amateur band.

Such an antenna also serves other needs for sundry emergency and service functions, including coverage of CAP and other frequencies close to the amateur band. The region below 2 meters is largely devoted to aeronautical mobile services, while the region above 2 meters is split between land and maritime mobile services. Let's see what an LPDA array to cover this wide frequency range looks like.

The Basic Design

Amateur-band LPDA design typically suffers from the attempt to use as few elements as possible on the shortest possible boom. LPDA design revolves around two mathematical variables: τ (*tau*), which defines the relationship between successive element spacings, and σ (*sigma*), the relative spacing constant. For every LPDA that uses less than the highest value of τ and the corresponding optimal value of σ, there will be only a few combinations that yield relatively high performance. For the present project, the design was restricted to 6 elements on a 54-inch boom, with a τ of 0.9238 and a σ of 0.1461. These values are short of the highest possible performance, but increasing τ would have required more elements and increasing σ would have lengthened the boom.[2]

For an LPDA, decreasing the characteristic impedance of the phasing line connecting the elements tends to increase array gain and to decrease the feed-point impedance. A 75-Ω phasing line yields acceptable 50-Ω performance from the array, with an average free-space gain across 2 meters of about 9.2 dBi. Since the front-to-back ratio of an LPDA tends to vary with the gain, it is uniformly high; that is, better than 30 dB across the band, with no strong rearward side lobes to decrease the overall front-to-rear ratio.

Figure 1 shows the general layout and basic dimensions of the LPDA that achieves this level of 2-meter performance. We shall examine a number of the construction details later in this article.

Of equal importance with the performance in the 2-meter band is how the array works across the entire operating passband. Figure 2 is a graph of the modeled free-space gain and front-to-back ratio of the LPDA from 130 to 170 MHz. The gain peak in the 2-meter band is readily apparent, with less but still useful gain above and below the desired design range. The front-to-back ratio only decreases below 20 dB above 160 MHz. The peak in the front-to-back curve is a normal LPDA phenomenon, since it peaks—often very sharply—at a frequency a bit lower than the peak frequency for gain.

For monitoring frequencies well outside the amateur band, these performance characteristics are quite serviceable. This LPDA was modeled with a 4-inch shorted transmission-line stub at the rear element. A stub between 2 and 4 inches long is necessary to avoid a pattern reversal at about 160 MHz, and it improves the overall performance of the array within the 2-meter band too. A stub length shorter than 4 inches will increase the 2-meter front-to-back ratio by about 2 dB, while reducing the operating range to a lower limit of about 132 MHz.

Figure 3 provides a look at the excur-

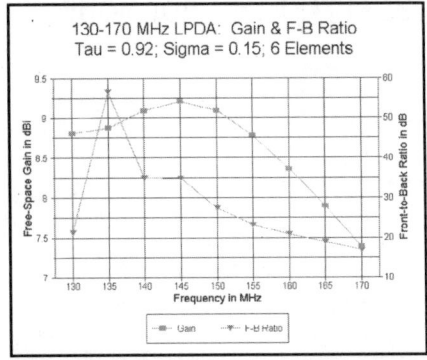

Figure 2—Modeled free-space gain and 180° front-to-back of the 2-meter-plus LPDA from 130 to 170 MHz.

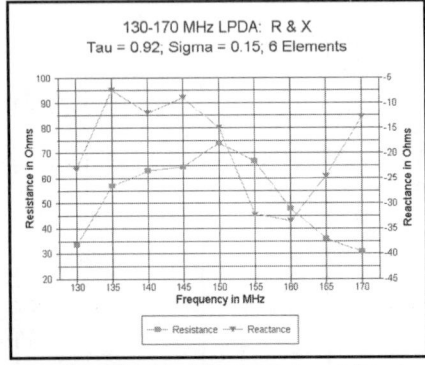

Figure 3—Modeled feed-point resistance and reactance of the 2-meter-plus LPDA from 130 to 170 MHz.

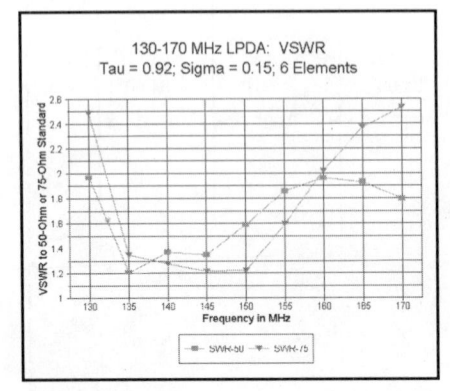

Figure 4—Modeled 50-Ω and 75-Ω VSWR of the 2-meter-plus LPDA from 130 to 170 MHz.

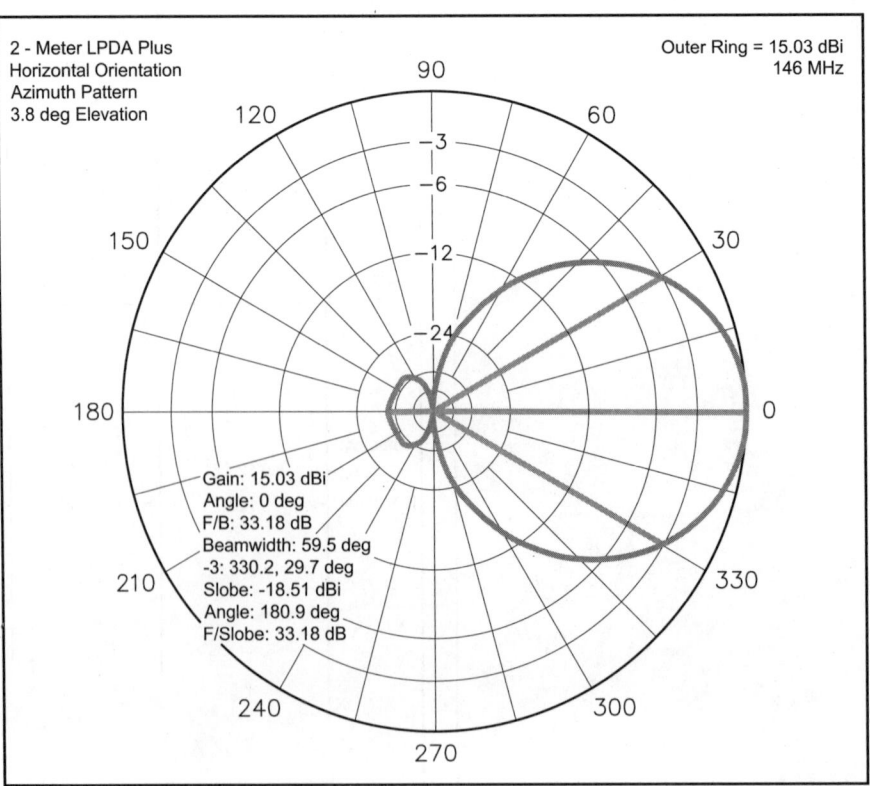

Figure 5—Free-space azimuth pattern of the 2-meter-plus LPDA at 146 MHz with elements oriented horizontally.

sions of feed-point resistance and reactance. The curves are quite normal for an LPDA. These arrays tend to show maximum and minimum values of resistance and reactance—as referenced to a median value for each—in a relatively *out-of-phase* pattern. The result is a broad-banded SWR curve when plotted against the median feed-point resistance value, in this case, about 60 Ω.

The 50-Ω and 75-Ω SWR curves appear in Figure 4. Within the 2-meter band, 75-Ω cable would be a slightly better choice of feed line—if you are looking for the lowest possible SWR. However, for the widest operating passband at an SWR of less than 2:1, 50-Ω cable is the better choice, since the resistive component of the feed-point impedance shows values well under 50 Ω at the high end of the operating passband.

Mounting the LPDA

You can orient an LPDA either vertically or horizontally. However, at conventional mounting heights, which are often (but not always) below 5 λ above ground, antenna far-field gain will vary according to orientation. Figure 5 shows the modeled far-field performance of the array at a height of 25 feet (300 inches or about 3.75 λ in the 2-meter band), with the antenna oriented horizontally. The operating frequency is 146 MHz. The 9.2 dBi free-space gain, when taking into account ground reflections, becomes just over 15 dBi at the elevation angle of maximum radiation (3.8°). Figure 5 also clearly shows the well-controlled rear pattern of an LPDA that uses high values for both τ and σ.

If we remodel the antenna to place the boom at the same height, but with the elements oriented vertically, we obtain the

The LPDA with elements oriented vertically. Use this orientation only with a nonconductive mast.

pattern in Figure 6. The −3 dB beamwidth has increased by about 25° for both the forward and rear patterns. The elevation angle of maximum radiation is 3.5°. The price of having a significantly wider beamwidth is forward gain, which is about 1.7 dB lower than the value shown in Figure 5 for the horizontal mounting configuration. An array must be well above 5 λ above ground before the gain figures for the two orientations begin to converge.

Questions arise from time to time about whether the far field patterns are good indicators of the antenna patterns in ground-wave point-to-point service. Figure 7 compares the relative patterns for the two orientations, using a receiving point 1 mile from the antenna at 25 feet above ground. It's clear that the antenna retains its pattern shape in point-to-point service. However, these patterns presume a clear field between the two antennas. Intervening objects and terrain variations can modify the actual performance of an antenna between any two stations.

Construction

Table 1 provides the basic dimensions of the LPDA array, in both inches and millimeters. The half-length values are important for construction, since each element is split at the center and connects to the phasing line.

The boom and phasing line for this design are one and the same. I chose $3/4 \times 3/4 \times 1/8$-inch thick aluminum channel stock for the twin-boom. This stock can be obtained from some hardware outlets and can often be special ordered if not immediately available.[3] The choice of thick-wall stock (in contrast to the same material with a $1/16$-inch wall thickness) arose from the element size and mounting detail I selected for the antenna.

U-channel has been used in a number of commercial antennas for VHF and UHF booms. Very often, commercial antennas will pressure-fit elements into the stock. For home shop construction, I use a different system. I picked $3/16$-inch diameter elements because they remain strong when the ends are threaded for 10-24 nuts. If I had tried to use $1/8$-inch diameter elements, they would be fragile when threaded. The selection of 10-24 threads required thick enough U-channel stock so that the threaded holes have enough threads to grab the element.

Figure 8 shows a cut-away end view of the scheme. I drilled $5/32$-inch holes in the two sides of the channel stock for each half element and then threaded them for 10-24 bolts. About $3/4$- to $7/8$-inch of the end of each half element is also threaded. As I screwed

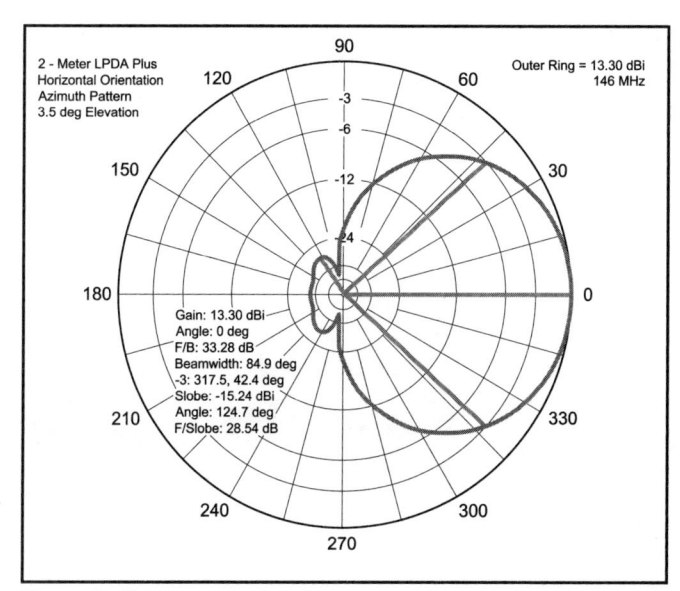

Figure 6—Free-space azimuth pattern of the 2-meter-plus LPDA at 146 MHz with elements oriented vertically.

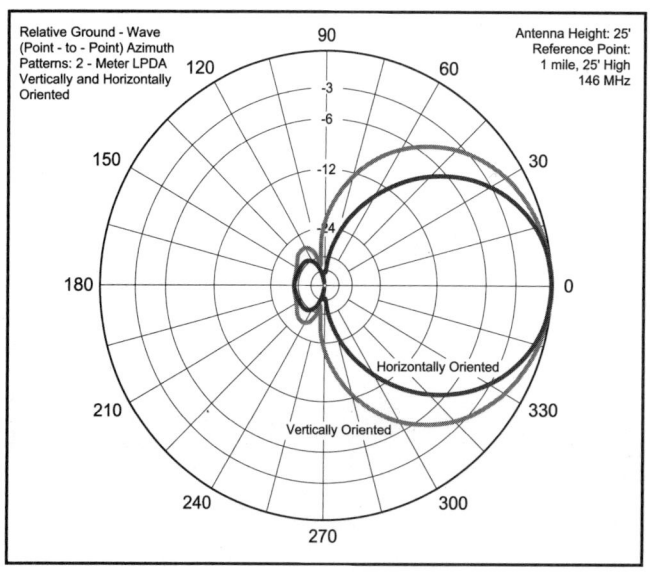

Figure 7—Relative ground-wave azimuth patterns with the elements oriented horizontally and vertically.

Table 1
2-Meter Plus LPDA Dimensions, in inches and mm

Ele #	inches				mm			
	Length inches	Half Length	Space from Ele n–1	Space from Rear Ele	Length mm	Half Length	Space from Ele n–1	Space from Rear Ele
1	43.02	21.51	—	—	1092.6	546.3	—	—
2	39.74	19.87	12.57	12.57	1009.4	504.7	319.4	319.4
3	36.72	18.36	11.62	24.19	932.6	466.3	295.1	614.5
4	33.92	16.96	10.73	34.92	861.6	430.8	272.6	887.1
5	31.34	15.67	9.92	44.84	796.0	398.0	251.9	1138.9
6	28.95	14.48	9.16	54.00	735.4	367.7	232.7	1371.6

the half element through the first side of the channel, I threaded two stainless steel 10-24 nuts onto it. I screwed the half-element end into the far channel wall until it just met the outer surface. Then I tightened the two nuts against the inner walls to lock the element.

Note that using this system requires that you add $^3/_8$ inch to each half-element length in Figure 1 and Table 1. The U-channel centerline is the reference point for all half-element lengths.

For my prototype, I used three nuts on the front half-elements, with a solder lug sandwiched between nuts on the element sides. I later soldered the coax cable to the feed-line lugs. The extra nuts on the rear element halves also do double duty when I added the shorted stub to the assembly.

The separation between U-channel faces is not at all arbitrary. The flat stock faces form a parallel transmission line. The use of flat-faced stock for the boom requires some adjustment when calculating the characteristic impedance of the phase-line. For conductors with a circular cross-section,

$$Z_0 = 120 \cosh^{-1}\frac{D}{d} \approx 276 \log_{10}\frac{2D}{d} \quad (Eq\ 1)$$

where D is the center-to-center spacing of the conductors and d is the outside diameter of each conductor, and D and d are in the same units of measurement. Since we are dealing with closely spaced conductors, relative to their diameters, the following adjustment to the equation for calculating the characteristic impedance (Z_0) yields more accurate results. For a square or flat-face conductor,

$$d^3 1.18\ w \quad (Eq\ 2)$$

where d is the approximate equivalent diameter of the square tubing or flat-faced stock and w is the width of the stock across the facing side.

For a given spacing, a square or flat-face stock permits you to achieve a lower characteristic impedance than with a round conductor. The approximation is useful but not precise, especially for stock that is not perfectly square. However, it is only necessary that the phasing-line impedance be close to 75 Ω to achieve the desired results with the present array. For $^3/_4$-inch U-channel, a spacing of about 0.32 inches (8.1 mm) is close enough for all practical purposes. The spacing can be adjusted during testing, with a closer spacing yielding—up to a point—a lower phasing-line impedance and feed-point impedance, with a potentially better 50-Ω SWR curve. However, too close a spacing (less than 0.25 inch or 6.3 mm) may be self-defeating, by altering array performance at one or the other ends of the operating passband.

Since the characteristic-impedance calculation presumes an air dielectric between conductors, I employed insulating spacers attached to the sides of the U-channel stock.

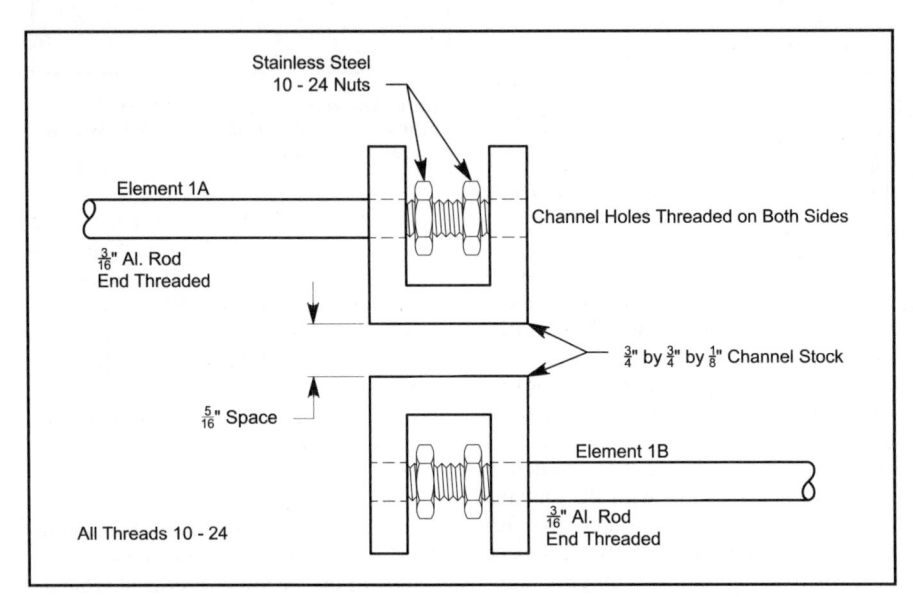

Figure 8—Cutaway end view of the twin-boom U-channel element mounting system.

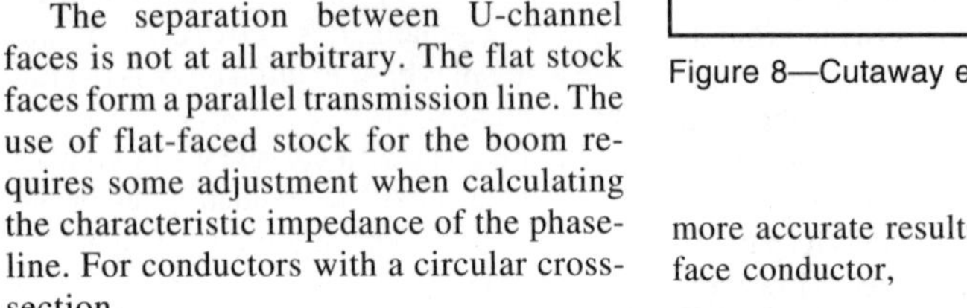

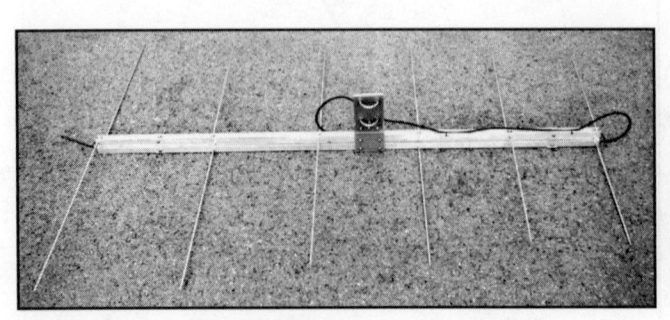

Side view of the array, showing the main mounting plate.

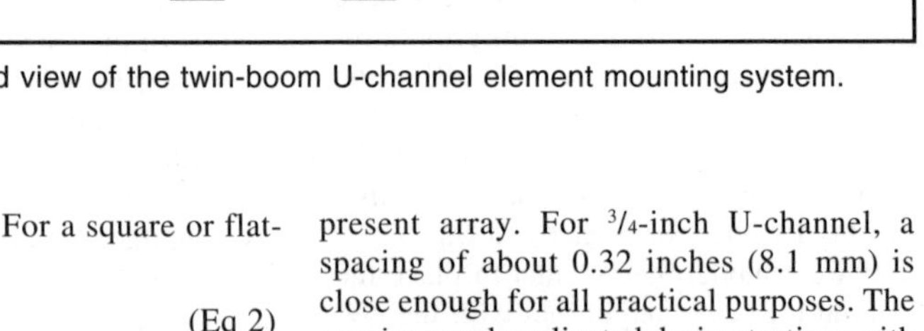

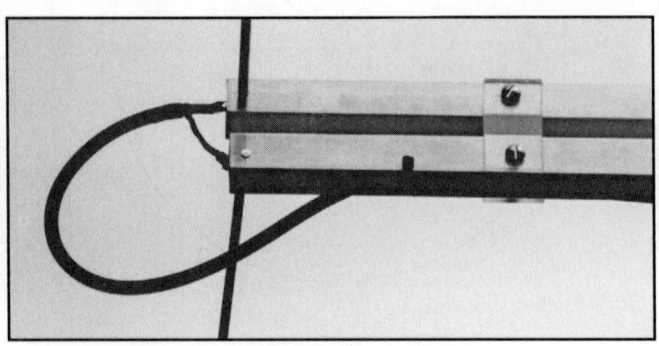

View of the front of the LPDA showing the elements, the feed-line mounting system and the front Plexiglas boom insulators.

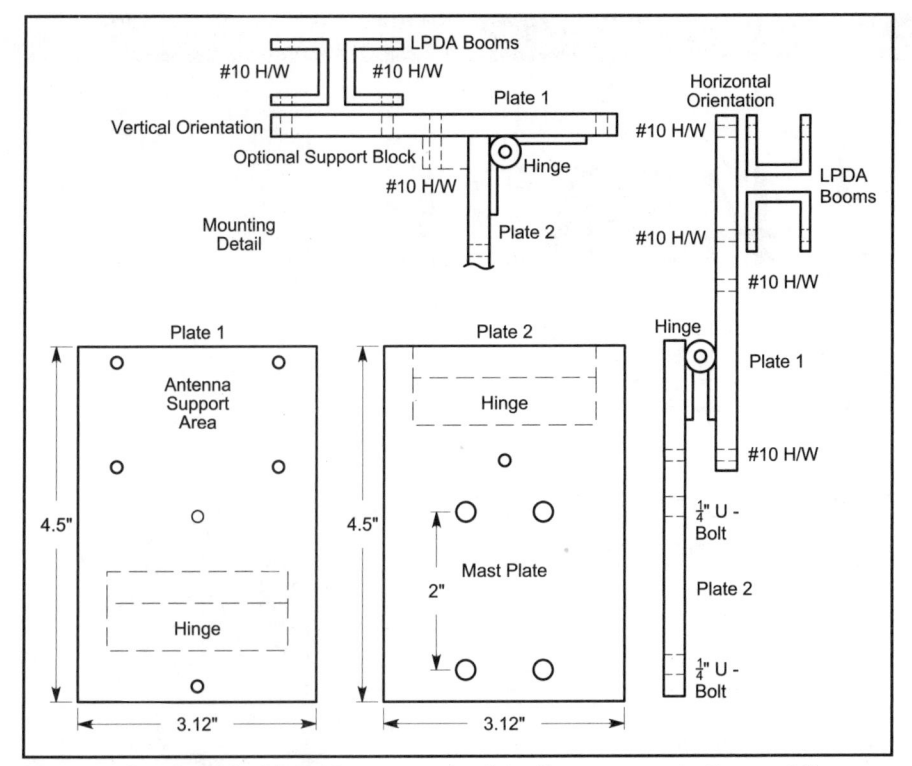

Figure 9—Details of the hinged boom-to-mast plate. Use a nonconductive mast if elements are oriented vertically.

Between the most forward element and the next one—and likewise between the rearmost element and the next one—I attached scrap Plexiglas strips on both sides of the twin boom. This is shown in the photo showing a side view of the array. At the boom center, I used a mounting plate with through bolts to support and separate the U-channel pieces.

The mounting-plate system was designed to permit the antenna to be oriented horizontally or vertically for various tests. A simpler system is certainly possible using a single mounting plate. The basic requirements are that the mounting system establishes the boom separation and that it holds fasteners (normally U-bolts) for attachment to the mast.

Figure 9 shows the double-plate system that I used for the prototype. Plate 1 holds the antenna's double boom at the approximate center point between elements 3 and 4. Stainless steel #10 nuts and bolts secure the boom to the plate. Plate 2 secures the assembly to the mast and is drilled for 1.25-inch wide stainless steel U-bolts. The most interesting feature of the mounting is the 2.5-inch hinge—rated for outdoor use—connecting the two plates.

The plate stock I used was 3/8-inch scrap fiberglass, which is structural overkill to some degree. Figure 9 shows the dimensions of the pieces that I used. When the antenna is vertically oriented, the top of antenna Plate 1 rests on the mast Plate 2 edge. For permanent use, I would add a further support epoxied and screwed to the mast plate. A stainless steel bolt would then lock down the antenna plate. When the antenna is horizontally oriented, the antenna plate is vertical and locked to the mast plate with a similar bolt.

The coaxial feed line for my prototype is RG-8X. A series of small holes (1/8 inch diameter) in the lower U-channel permits the cable to ride inside cable tie loops within the channel until it reaches the mast area. At the forward or feed end of the array, the cable center conductor and braid connect to the previously mentioned solder lugs attached to each side of the front element. These connections need to be weatherproofed by a suitable *cap* structure or by applying standard weatherproofing techniques. For the prototype, the cable end was coated with a plastic dipping compound available at home centers. Since it returns under the lower U-channel, you should size the coax loop to avoid internal deformation as the weather changes from cold to hot and back and to avoid stress on the connection points.

The final step in the process is to add the shorted stub to the rear of the array. I used a length of 75-Ω cable (RG-59U) with foam insulation, with a velocity factor of about 0.78. Hence, my 4-inch stub is about 3.1 inches long physically to account for the approximate 0.78 velocity factor of the line. Like the feed line, the ends of the cable were dipped in a plastic compound to provide a weather-seal.

In the final version, you might simply extend the twin booms 4 inches to the rear of the last element and connect the boom ends. This system might require moving the mounting plate to the rear slightly to keep the weight reasonably balanced. The array is likely a bit too heavy for effective rear-end mounting.

The stub completes construction. Lighter construction is undoubtedly possible, since the weight of the stock used in the prototype more nearly approximates commercial service sizing. However, the antenna has stood up to rough use in testing.

Performance

The photographs show the finished product, which I tested at 6, 10 and 15 feet above ground. The view of the antenna showing the elements oriented vertically is for photographic purposes only. The use of a metal mast would actually detune the array. When I changed the upper 5 feet of mast to a length of PVC, everything returned to normal—that is, to the values obtained with the elements horizontally oriented.

Initial SWR curves were taken with 20 feet of RG-8X between the array and an MFJ 259B analyzer, frequency calibrated to a 2-meter receiver. Within the 2-meter band, measured SWR was 1.5:1 or better. The predicted 2:1 SWR curve for the model (Figure 4), which did not employ a feed line, ranged from 130 to 170 MHz. The measured SWR provided less than a 2:1 SWR from 124 to 172 MHz. Part of this frequency range expansion is due to cable losses. However, the greater low-end extension of the curve suggested that the stub might be a bit long relative to the 4-inch equivalence desired. A ruler confirmed the suspicion, since the stub lead lengths had not been fully accounted for during construction.

Although I have no antenna range on which to directly confirm gain and front-to-back values, the array gain equaled that of other antennas in my shop of similar capabilities. With the antenna vertically oriented, I was able to silence all but one local repeater for over 180° of array rotation, indicating that the front-to-back ratio was as modeled. With a borrowed scanner, I received numerous signals at full quieting throughout the design passband.

The LPDA described here is not a competitor to wide-band Yagis designed expressly for 2 meters. Instead, it is a complementary antenna, designed for good 2-meter performance, but with additional capabilities over the 130-170 MHz range. If the wider-band service of an antenna is among your needs, then this 2-meter-plus LPDA may find a niche in your gallery of antennas.

Notes

[1]See the article "In Pursuit of Better VHF Quad Beams: A Work in Progress" in the 2001 *Proceedings of the Southeast VHF Society* for details of a wide-band Yagi meeting the specifications noted in the text. An alternative but close set of dimensions is provided in an article at my Web site (**www.cebik .com**) in the item called "High-Gain, Wide- Band Yagis for 10, 6 and 2 Meters." This item first appeared in *AntenneX*, Aug 1999 (**www.antennex.com**).
[2]See Chapter 10 of *The ARRL Antenna Book*, 19th Edition, for a full explanation of LPDA design and the fundamental design factors, τ and σ.
[3]My thanks to Raul Pla, W4AWI, of Antenna World, who generously donated the U-channel stock for this project.

A 146- and 445-MHz J-Pole Antenna

Pipe your signals to 146 and 445 MHz with one antenna!

Getting on 146 and 445-MHz with a single J-pole antenna can be done inexpensively. I did it by building the dual-band J pole shown in Figure 1. The total materials cost about $21, and only commonly available hand tools are required for assembly. Interested?

Some Background

A vertical J-pole or dipole designed for use at 146 MHz will resonate at 440 MHz because it's about $3/2\lambda$ long at that frequency. However, according to *EZNEC*,[1] most of the 445-MHz radiation is at an elevation angle of about 45° instead of a lower angle desired for repeater and ground-wave communication. Also, the antenna's input impedance at 445 MHz is about two and a half times that of the 146-MHz value. For dual-band operation, both of these hurdles can be overcome by simply placing two 445-MHz elements close to the feedpoint of the 146-MHz $1/2$-λ element. The vertical radiation pattern of the resulting antenna at 445 MHz is shown in Figure 2. The 445-MHz elements have little effect on 2-meter operation. Once the antenna is adjusted for 2-meter operation, the 445-MHz antenna input impedance can be adjusted to equal the 2-meter impedance by adjusting the spacing of the 445-MHz elements from the main element. Increasing the spacing between the elements increases the impedance at 445 MHz and vice versa. The length of the 445-MHz elements primarily affects the resonant frequency and to some extent, also affects the input impedance. The length of each 445-MHz element is less than $1/2\lambda$ at 445 MHz. At a true $1/2\lambda$, the impedance and resonant frequency appear to be insensitive to spacing and length adjustments.

At 146 MHz, this antenna's input impedance is about 65 Ω, delivering an SWR of about 1.3:1 at resonance. Placing a $1/4$-λ Q section in the feed line at the feedpoint can lower the SWR between 144 and 148 MHz. The Q-section impedance is about 59 Ω. Because the Q-section length is about $3/4\lambda$ at 445 MHz, it also works at this frequency.

How It Works

The antenna's main vertical element (see Figure 1 and the title page photo) is about $1/2\lambda$ long at 146 MHz and employs a $1/4$-λ stub at the bottom to decouple the main element from the mast and feed line. The antenna is similar to a standard J-pole antenna [2,3] except that it's *fed at the center of the main element* instead of tapping the feed line partway up the stub. The coaxial-cable feed line passes through the main element. Two elements, almost $1/2\lambda$ long at 445 MHz, are placed near the antenna's feedpoint and parallel to the main element. These elements are parasitic and don't need a separate feed line; they are excited by the main vertical element. The antenna is quite efficient because no lossy matching networks or coils are used. The gain on both bands is about the same as a vertical dipole or single-band J pole.

As described later, the Q-section is made by replacing $13^{5}/_{8}$ inches of the coaxial-cable feed line shield braid with $13^{5}/_{8}$ inches of $3/8$-inch copper tubing.

Construction

The antenna elements are made of $1/2$-inch copper water pipe and soldered fittings. The center insulator (see Figure 3) is made from $1/2$-inch PVC pipe and the standoff insulators (see Figure 4) for the 445-MHz elements are made of $1/4$-inch-thick Plexiglas. You can purchase the $1/2$-inch copper pipe, copper fittings and $1/2$-inch PVC pipe from a building supply outlet. (If you're going to build two antennas, you can purchase the copper pipe much cheaper in 20-foot lengths at a plumbing supply.) The Teflon-silver PL-259 connector and RG-8 coax are available from RadioShack. (You can find good buys on such connectors at hamfests.) Short lengths of $3/8$-inch-diameter tubing can be purchased at a hardware store. Small pieces of $1/4$-inch-thick Plexiglas can usually be found as scrap at a glass shop. Broken golf-cart windshields are another source of Plexiglas.

In any 445-MHz antenna construction project, it's important to adhere to the given dimensions. A dimension deviation of even $1/16$ inch is considerable at 445 MHz, especially at the feedpoint. It's practically

Table 1
Materials Required

10 feet of copper water pipe (type L or M; see text)
5 inches of $1/2$ inch PVC pipe
1—$1/2$ inch copper T
1— $1/2$ inch copper L
2—$1/2$ inch copper pipe caps
15 inches of $3/8$-inch OD copper tubing (0.331-inch ID)
2—Pieces of $1^{1}/_{4}$ in × $6^{1}/_{8}$ in × $1/4$-inch-thick Plexiglas
5 ft of RadioShack RG-8 coax (RS 278-1312)
1—Teflon-silver PL-259 connector.

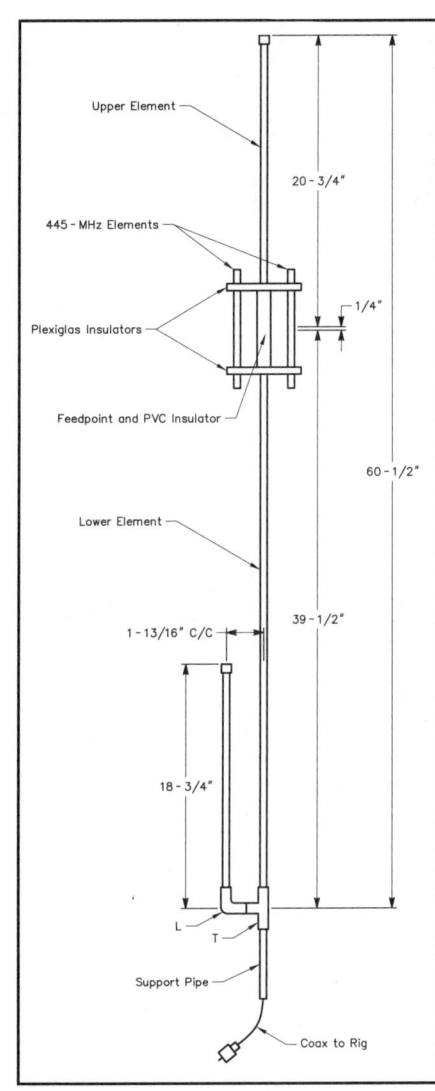

Figure 1—General dimensions of the two-band J-pole antenna. The copper-plumbing **L** and **T** at the bottom of the antenna fix the spacing between the stub and lower portion of the main element.

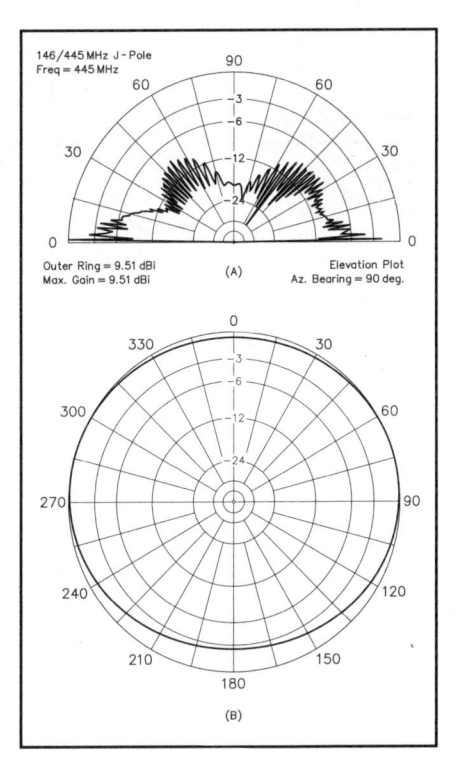

Figure 2—*EZNEC* plot of the antenna's vertical (A) and azimuthal (B) radiation patterns at 445 MHz.

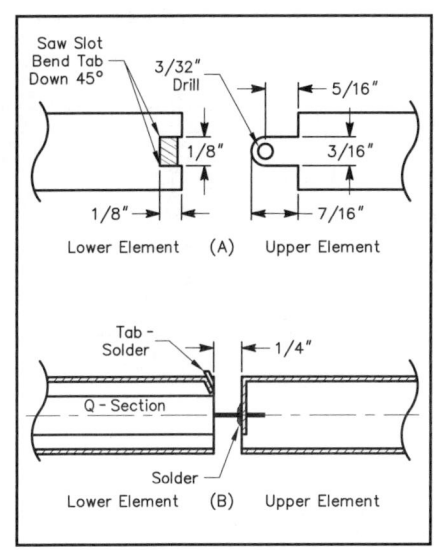

Figure 3—Feedpoint detail. The tab on the upper portion of the main element accepts the center conductor of the coax of the **Q** section. See Figure 9.

impossible to construct a feedpoint connection at 445 MHz without introducing some transformation of the antenna impedance. Therefore, you should closely follow the dimensions and feedpoint detail described. I built a second antenna using the plans provided here and it performs exactly like the prototype. If you intend to mount the antenna on a mast, use type L copper pipe for the elements. Use type M pipe (it's lighter than type L) if you intend to suspend the antenna from a support.

Before soldering, polish all mating pipe pieces with #0000 steel wool. I recommend using a propane torch for soldering the joints. The trick in soldering copper pipe and fittings is to get the copper hot enough to melt the solder *before* applying solder. The solder then flows into the joint without leaving drips that require cleanup.

First, cut two pieces of copper pipe to make the stub and the 39$\frac{1}{2}$ inch lower element. Cut these pieces to their final lengths after soldering each piece to the respective

L and **T** at what will be the bottom ends of the stub and lower element. Measure the length of these pieces from the *center* of the **T** and **L** connection. Cut the stub pipe to a length of 18$^{11}/_{16}$ inches allowing about $^{1}/_{16}$ inch for placement of a cap at the top of the stub. Use a 1-inch length of pipe to join the **T** and **L**. The **T** and **L** butt together to fix the 1$^{13}/_{16}$ inch center-to-center spacing between the stub and the lower element. The mounting/support section below the lower element can be any length, but should be at least 12 inches to allow clamping to a mast.

Make two hacksaw slots in the top of the lower element as shown at the left in Figure 3A. Bend the tab between the slots about 45° toward the center of the pipe.

Before cutting the upper element to length, some work must be done. At what will be the bottom of the upper element, cut a tab as shown at the right in Figure 3A. Drill a $^{3}/_{32}$-inch diameter hole through the end of the tab. (It is at this tab where the center conductor of the feed-line coax will later be attached.) Dress the tab with a file and tin the tab using a propane torch or high-power soldering iron. Bend the tab 90° toward the pipe center as in Figure 3B. *Once the tab is bent,* cut the upper element to a length of 20$^{3}/_{4}$ inches allowing about $^{1}/_{16}$ inch for placement of a cap on the upper-element top.

You can fabricate the standoff insulators shown in Figure 4 from $^{1}/_{4}$-inch-thick Plexiglas or $^{1}/_{2}$-inch PVC pipe. Plexiglas is easier to use because the $^{5}/_{8}$-inch-diameter holes can be made using a hand-held drill and a common wood spade bit. Drill at a low speed to prevent melting the Plexiglas.

If you use PVC pipe for the standoffs, use a drill press to keep the holes properly aligned. Cut the two 445-MHz elements to a length of 11$^{1}/_{16}$ inches (see Figure 5). Make the center insulator from a 5-inch length of $^{1}/_{2}$-inch PVC pipe (see Figure 6). Cut a longitudinal slit the entire length of the pipe; I used a hacksaw to do this.

The coaxial-cable feed line extending from the antenna's feedpoint to just below the support section can be any convenient length, but use RadioShack RG-8 (RS 278-1312)[4] to get the proper velocity factor and impedance for the Q section. The Q-section details are shown in Figure 7. The Q-section consists of a 13$^{5}/_{8}$-inch length of $^{3}/_{8}$-inch copper tubing (Figure 7A) slid over the center insulation of the top end of the feed line (Figure 7B).

On the Q section, fashion a tab on the end of the tubing similar to that at the bottom of the upper element (see Figure 7A). Following Figure 7B, cut the end of the feed line with a sharp knife to expose $^{7}/_{16}$ inch of the center conductor. (Figure 8 shows the finished feedpoint end of the Q section.) Remove 13$^{1}/_{4}$ inches of the coaxial-cable feed line's outer cover and the shield to expose the dielectric. Then remove an additional $^{5}/_{8}$ inch of the coaxial cable's outer covering exposing the shield braid. Clean and tin the tab and the opposite end of the Q-section tubing. Slide the tubing over the coax and under the shield so that the end of the tubing with the tab is even with the end of the center dielectric (Figure 7B). Using a high-power soldering iron or low torch flame, solder the shield to the tubing, allowing the solder flow through the braid. Dress the joint so that it passes through the lower element.

Assembly

Clamp the lower element and stub assembly horizontally in a vise about 14 inches

Multiband   5-7

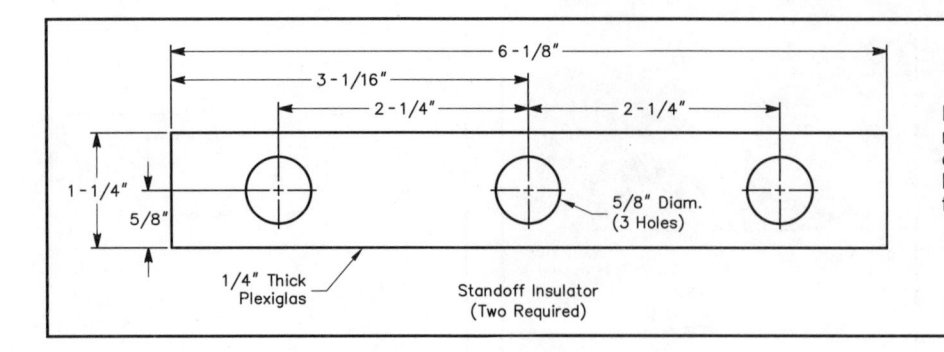

Figure 4—Two standoff insulators made of ¼-inch-thick Plexiglas sheet or ½-inch PVC pipe are required to hold the 445-MHz elements adjacent to the main element.

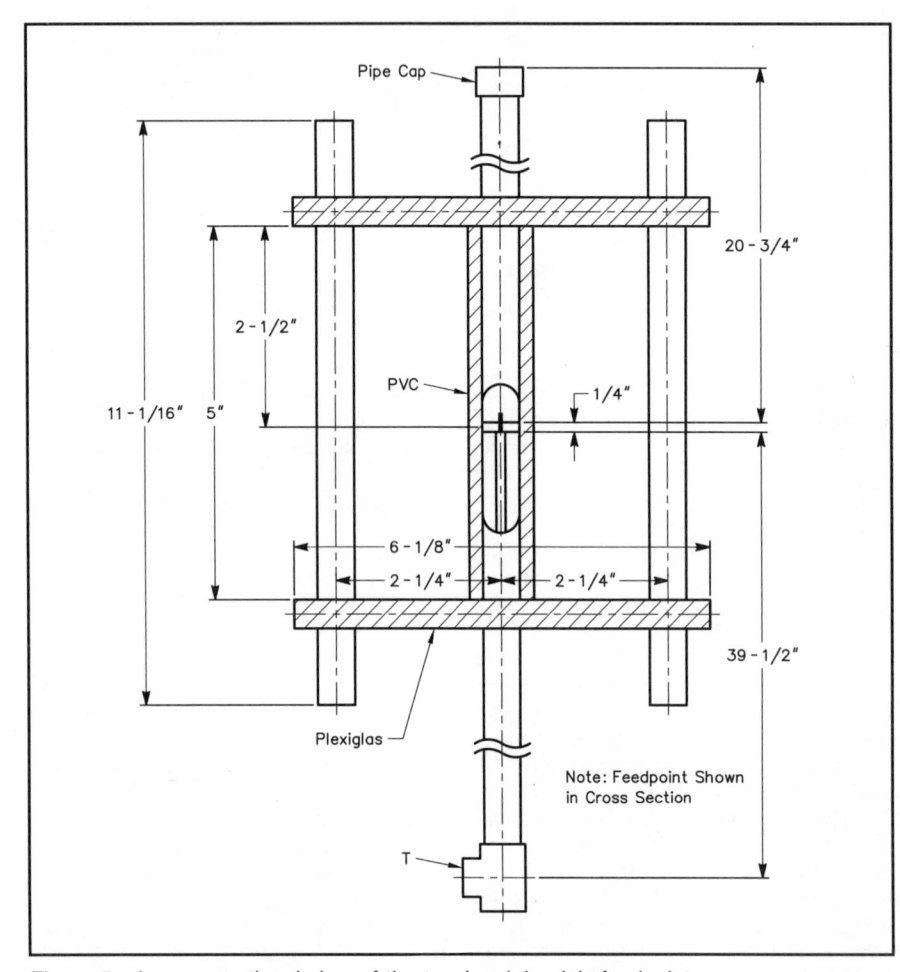

Figure 5—A cross-sectional view of the two-band J-pole's feedpoint.

from the top of the lower element. Slide the Q section and feed line into and through the support section until the Q section is exactly even with the top of the lower element. While keeping the Q section in the center of the lower element (Figure 3B), bend the tab on the Q section over the tab on the lower element. Solder the tabs together. Snip off the excess tab length of the Q-section and file the joint flush with the surface of the lower element.

Slide one standoff onto the lower element until it is about 10 inches from the feedpoint. Open the slit in the center insulator and hold open the slit by inserting two pennies (or something similar) side by side at four points within the slit. Slide the center insulator over the lower element until it touches the standoff. Bring the upper element into place passing the center conductor of the coax into the hole in the tab according to Figure 3B. To hold the feedpoint in place, clamp the lower and upper elements to a strip of wood using hose clamps. Solder the coax center conductor to the upper-element tab by heating the upper element near the base of the tab and letting the heat flow down the tab.

Remove the hose clamps and wood strip around the feedpoint. Slide the center insulator up until it is centered over the feedpoint and remove the pennies. The center insulator will clamp tightly around the feedpoint. As shown in Figure 5, assemble the 445-MHz elements and upper standoff. Use silicone cement or caulking to hold the 445-MHz elements in place. Seal the top,

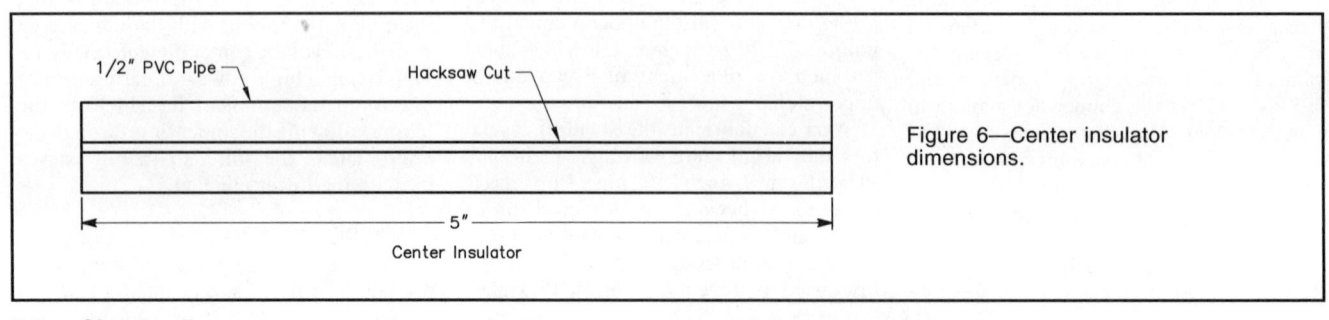

Figure 6—Center insulator dimensions.

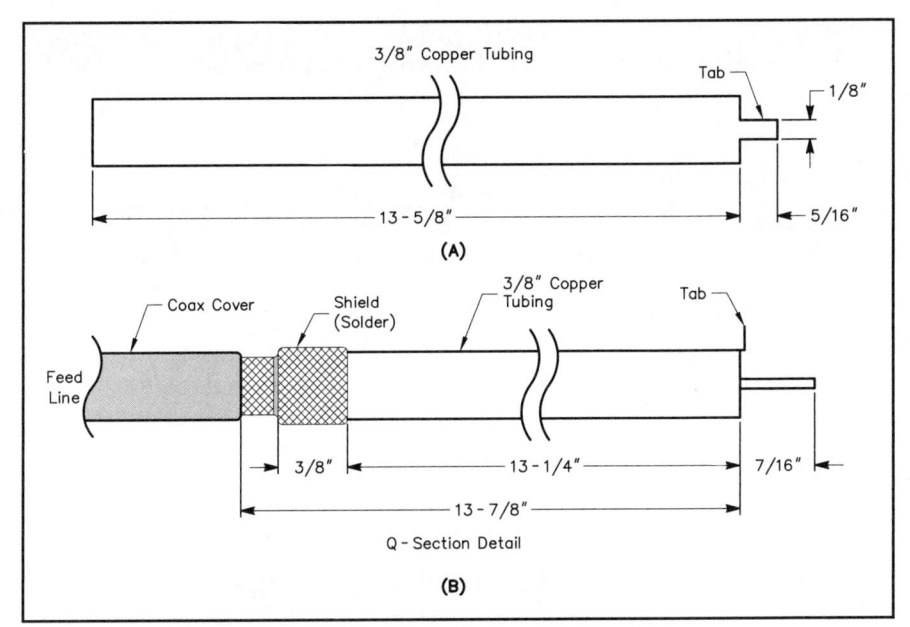

Figure 7—Q-section detail. A length of copper tubing is slid over the coaxial cable's dielectric and soldered to the shield braid at one end; see text.

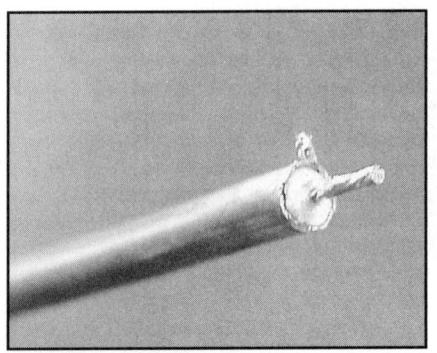

Figure 8—Feedpoint end of the Q section prior to final assembly.

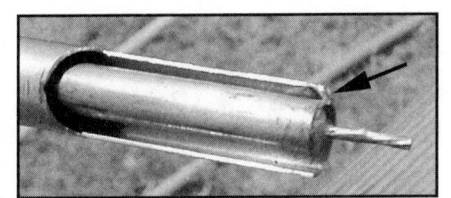

Figure 9—This cutaway view shows the tabs of the Q section and lower element soldered together.

bottom and slit of the center insulator with silicone cement or caulking. Install the cap on the top of the upper element.

If you intend to suspend the antenna by its top in a tree, install a top insulator made from PVC pipe and provide a support for the feed line at the bottom of the support section.

Evaluation

I found that the SWR and frequency range of the antenna are about the same when the antenna is mounted on a 20-foot mast as when it's suspended by its top at 40 feet. With 50 feet of CQ-4XL cable[5] ("poor man's Hardline"), the SWR measured 1.3:1 or less from 144 MHz to 148 MHz and 1.5:1 or less from 438 MHz to 450 MHz. Because I don't have a tower, I suspended the antenna in a tall pine tree. With the top at about 40 feet, I got the expected signal reports on 2 meters from repeaters within a 40-mile radius. The only repeater available to me in the 440-MHz band is about 30 miles away and at an elevation of only 100 feet, with its antenna on the side of the tower opposite my location. Six of seven bars on my LCD S meter lit up and I received good reports from stations that I worked. I've concluded that the radiation patterns and gain predicted by *EZNEC* are close to those realized on the air.

Notes

[1]*EZNEC* antenna software is available from Roy Lewallen, W7EL, PO Box 6658, Beaverton, OR 97007; tel 503-646-2885; fax 503-671-9046; **w7el@teleport.com**. Price: $89 postpaid; add $3 outside the US and Canada. Visa and MasterCard charge cards accepted.

[2]John Post, KE7AX, "The Copper Cactus J-Pole," *73 Amateur Radio Today*, Feb 1992, pp 9, 10 and 27.

[3]*The ARRL Handbook*, 1998 Ed, pp 20.56-20.57.

[4]RadioShack RG-8 cable has the following characteristics: a solid dielectric of 0.280-inch OD, a 13-gauge center conductor (7×21) and a velocity factyor of 0.66.

[5]The Wireman, 261 Pittman Rd, Landrum, SC 29356-9544; tel 800-727-9473 (orders only), technical assistance 864-895-4195; fax 803-895-5811; e-mail **cqwire@juno.com, n8ug@juno.com**; **http://thewireman.com**.

By Robin Rumbolt, WA4TEM
From *QST*, March 1993

An Easy, On-Glass Antenna with Multiband Capability

Are you mobile on one VHF/UHF band, or two? Either way, this on-glass antenna design is for you!

With every new car purchase comes the agonizing decision of where to punch the hole for my 2-meter antenna. Recently I purchased a dual-band transceiver, and the problem became where to punch two holes. I'd rather punch no holes at all!

An on-glass antenna seemed like the ideal solution. Such antennas couple RF through the windshield without the need to drill holes for cables and mounting hardware. Being a builder at heart, I designed an on-glass antenna to suit my needs. Not only does it feature the ability to disconnect the radiating element quickly (for car washes, etc), it has multiband capability, too!

Construction

I built the base of my antenna out of heatsink material (see Figure 1). I happened to find a piece of bare aluminum heat-sink stock with long, straight fins. Each fin was spaced about 1/4 inch apart. You can find similar heat-sink material at your local hamfest flea market. It's cheap and relatively easy to machine. You can also use aluminum channel stock, which is available from a variety of sources.

The first step was to cut out a piece roughly 1-5/8 inches square and remove all but the two middle fins. The fins were a bit too tall, so I carefully trimmed them to 1/2 inch in height. I used a grinding wheel to round the corners and drilled 9/64-inch holes in the centers of both fins.

The antenna coupling plate is cut from a piece of sheet steel. Its dimensions equal those of the antenna mount. (Avoid using aluminum for the coupling plate, since it's very difficult to solder.)

The quick-disconnect assembly is made from two hexagonal brass standoffs just wide enough to fit snugly between the fins. One standoff has a hole threaded through its entire length. The other standoff has a threaded stub on one end and a threaded hole in the other. I carefully drilled a 9/64-inch hole through the open end of the sec-

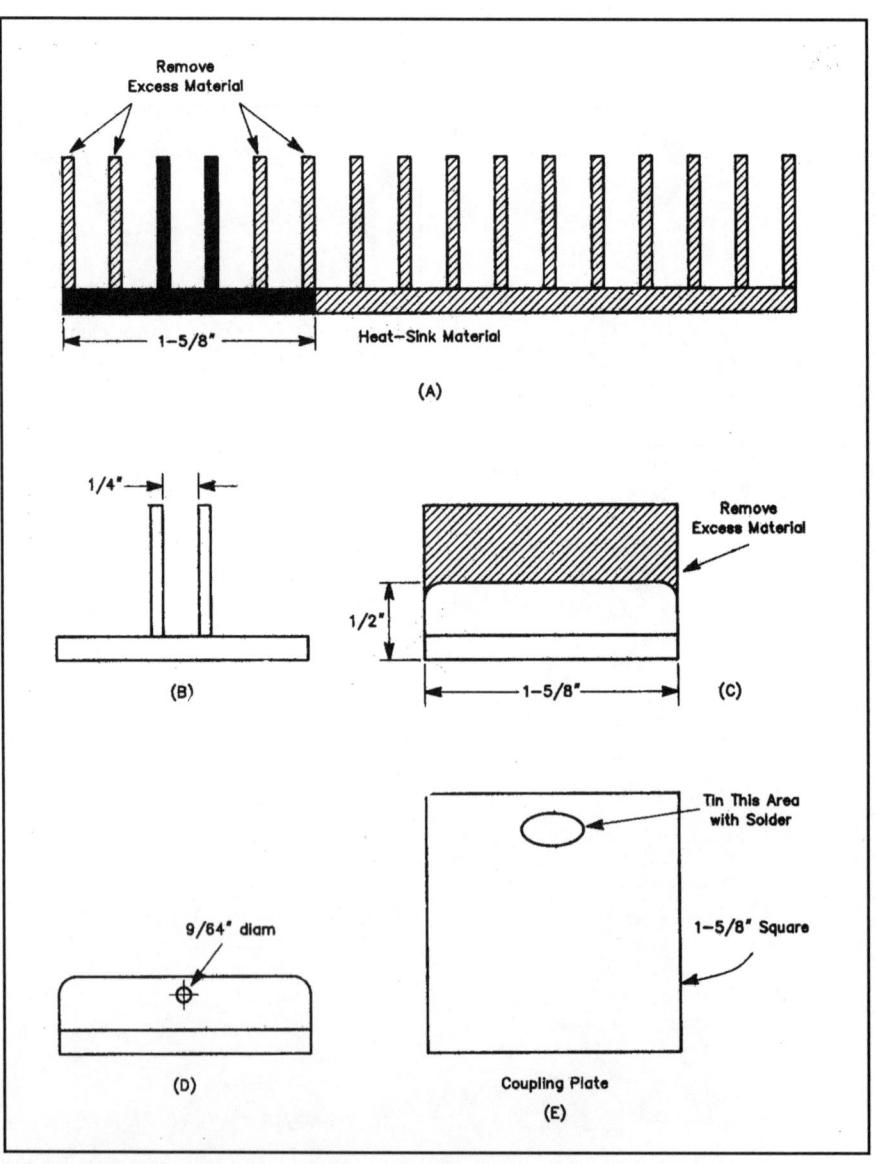

Figure 1—The antenna base is fabricated from a piece of heat-sink stock. Cut out a 1-5/8-inch section and remove all but the two center fins (A and B). Round off the sharp corners of the fins and trim for a 1/2-inch height (C). Drill a 9/64-inch hole through the centers of both fins (D). The coupling plate is cut from a 1-5/8-inch section of sheet steel. Tin a small area as shown at E.

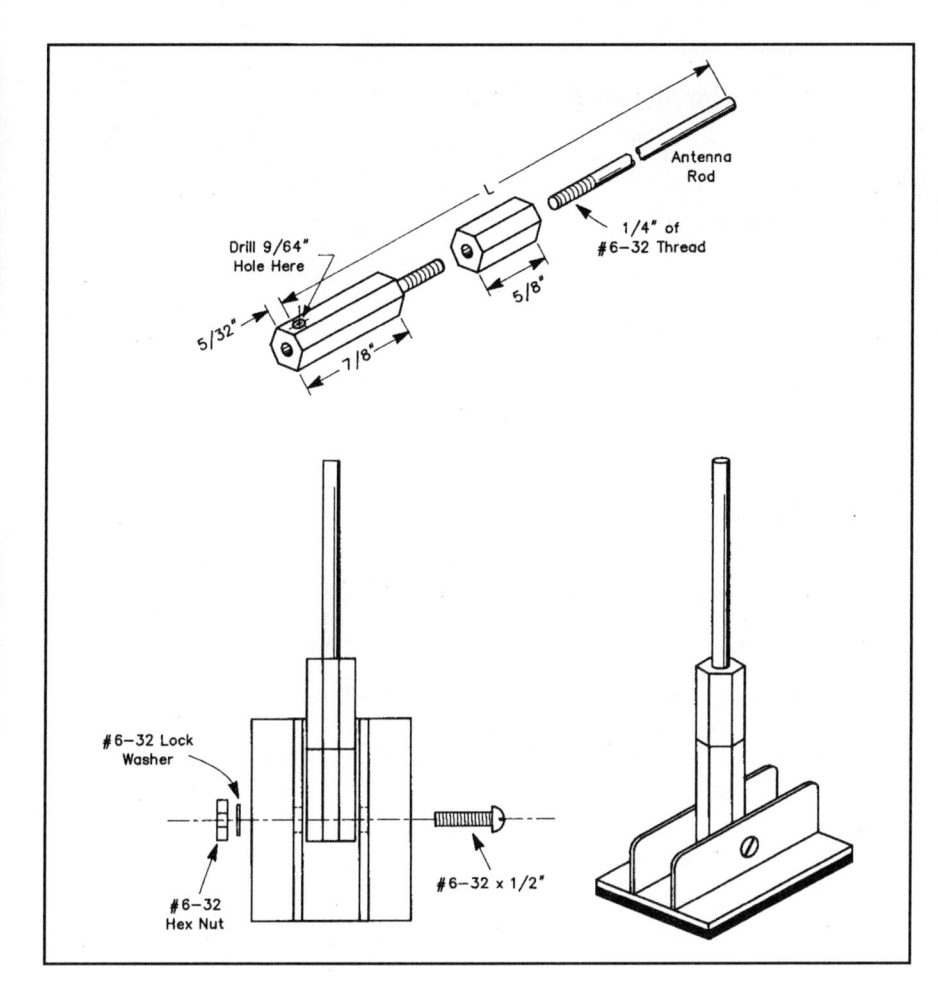

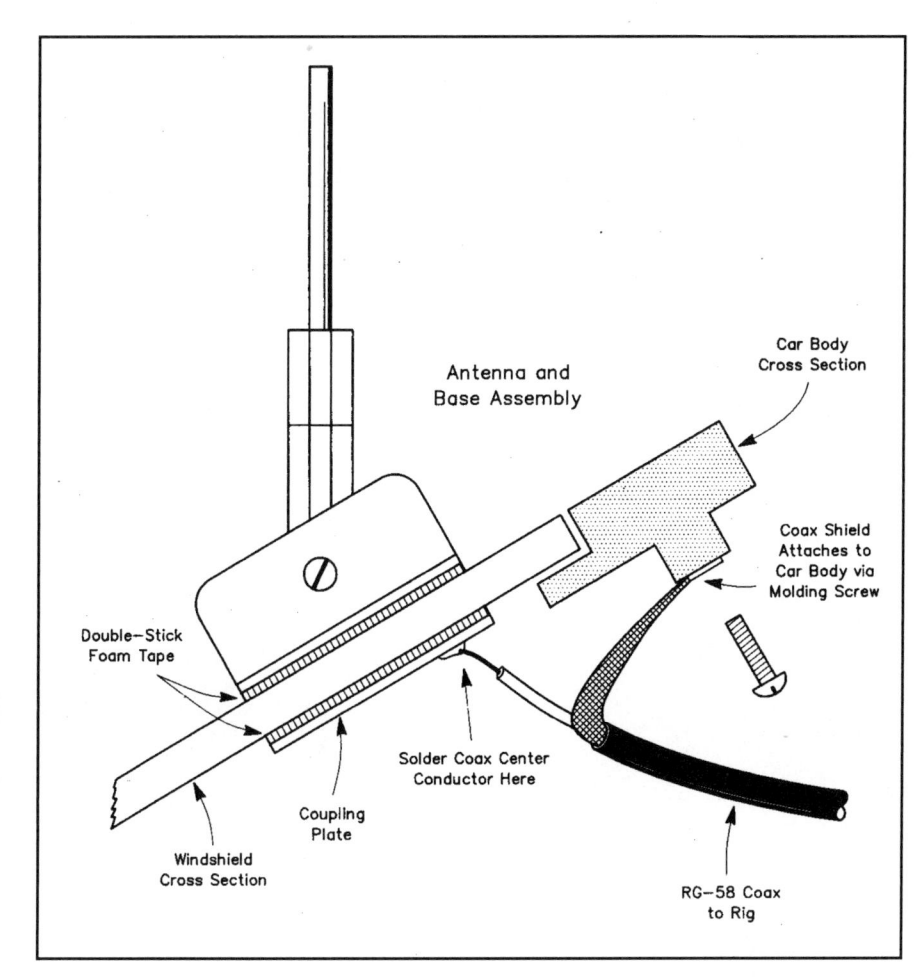

Figure 2—The on-glass antenna is made of brass welding rod attached to two threaded standoffs (see text). Drill a #6-32 hole in the bottom standoff as shown. Using a #6-32 × ½-inch screw, nut and lock washer, secure the antenna to the base. The total length of the antenna (L) is measured from the tip of the welding rod to the mounting screw. Use the lengths listed in Table 1 and then trim as necessary to obtain a low SWR.

ond standoff. Using a #6-32 × ½-inch screw, I assembled the standoff to the base as shown (see Figure 2).

The radiating element is made of 3/32-inch brass welding rod. I cut a #6-32 thread about ¼ inch up one end. This end is screwed tightly onto the first standoff. If you lack the tools to thread the rod yourself, use 1/16-inch welding rod and solder it to the standoff.

The total antenna length depends on the band you wish to use. See Table 1 for approximate lengths for various bands. As you can see in Figure 2, the finished section screws onto the stub of the base-mounted standoff. Whenever I need to remove it, a few twists is all it takes!

The coupling plate and the antenna base are attached to the windshield with double-sided foam tape (Radio Shack 64-2361). One tape strip isn't wide enough to cover the base and the plate, so I applied two strips side-by-side. It was a simple matter to cut the strips, peel off the backing and apply the tape to each piece. Any excess is easily trimmed away. The important thing to remember is not to peel the paper backing from the tape until just before you're ready to install the antenna.

Mounting

As you search for just the right spot to mount your antenna, bear in mind that you must ground the coaxial cable shield to the car body near the mount. In most cars, the top center of the front or rear windshields is best. Older cars usually have screws to attach the molding in these areas. These screws can often be used for grounding. If you own a newer car without strategically located screws, you'll have to install one yourself. In my Dodge Caravan, I drilled a small hole in the roof support (not the roof itself!) and used a small sheet-metal screw to fasten a solder lug in place. Whichever approach you use, check the screw with your VOM and ensure that it really makes contact with the chassis of your car. Many screws anchor in metal, but the metal isn't always grounded!

Figure 3—Use strong foam tape to hold the base and the coupling plate to the windshield. The coaxial cable center conductor is soldered to the coupling plate. The braid is grounded to the car body via a nearby molding screw. The braid must be grounded at the antenna for proper performance.

Multiband 5-11

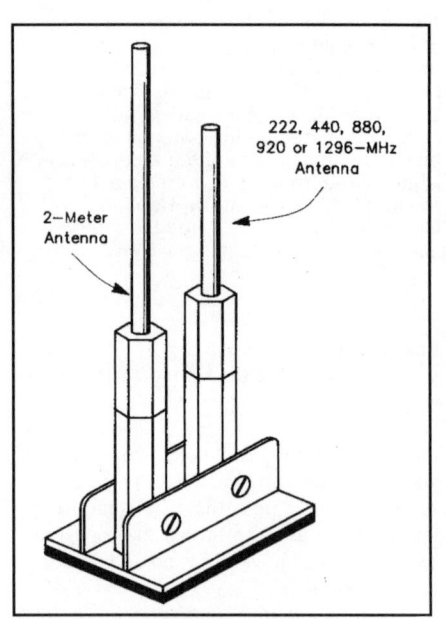

Figure 4—The multiband option. You can mount two antennas in the same base. This is ideal for today's dual-band, VHF/UHF transceivers.

Table 1
Antenna Lengths for Various Bands

Frequency (MHz)	Length (inches)
145	23¼
223	15¹/₁₆
444	7¹⁷/₃₂
880 (cellular telephone)	11³/₈ (¾ λ)
920	11 (¾ λ)
1296	7¾ (¾ λ)

Hold the base to your windshield in the area where you intend to install it. Adjust the antenna until it is vertical, then tighten the nut. Remove the mount and spray paint the entire assembly black—or whatever color looks best with your car!

When the paint is dry, clean the glass thoroughly (inside and outside). Check your chosen antenna location one more time. Is it in the path of windshield wipers? If you open the trunk or hatch, will the antenna be crushed?

If everything looks safe, peel the paper from the foam tape and attach the base to the outside glass. Press firmly to ensure that the tape sticks to the surface. Attach the coupling plate to the inside glass directly opposite the base. Solder the center coax conductor to the coupling plate and connect the coax shield to the ground screw or lug.

Tuning

With an accurate SWR/power meter, make SWR measurements and begin pruning the antenna for the lowest SWR. In my 2-meter installation, with 50 watts forward power, the needle doesn't even wiggle in the reverse-power position (a 1:1 SWR). If you can't get the SWR below 3:1, check your coax ground at the antenna. This is often the culprit.

Multibanding

I saved the best for last! You can create a dual-band antenna by simply mounting a second antenna and quick-disconnect assembly on the same mounting base (see Figure 4). For example, here's a fancy system for hams who own cellular phones: Install 2-meter and 880-MHz antennas in the same mount. The 2-meter whip will do double duty on 2 meters and 70 cm, while the 880-MHz antenna is perfect for your cellular telephone. An antenna farm on glass! (This configuration must be fed with a single feed line and an appropriate diplexer must be purchased or homebrewed.[1])

Conclusion

I am extremely happy with the antennas I have made using this on-glass method. No external holes were necessary and the antennas disconnect easily. SWR is low on every band and the antenna's radiation efficiency seems to rival any hole-mounted antenna I've used in the past!

[1] D. Jenkins, "A Simple VHF/UHF Diplexer," *QST*, October 1991, pp 18-25.

By J.L. Harris, WD4KGD

From *QST*, February 1980

A VHF/UHF 3-Band Mobile Antenna

Three bands—144, 220 and 440—on one stick sound interesting? This antenna might allow you to condense that stainless-steel and plastic jungle atop your auto onto a single pole.

In looking for a mobile antenna system for my Drake UV-3, I rejected the notion of one broadband antenna such as the discone because of band-switching problems not to mention its somewhat busy appearance. I also rejected the idea of three separate whips which I felt would give the relatively small roof area of my pickup truck a cluttered look. Three separate antennas confined to so small a space would also cast "shadows" on the vertical patterns of one another. In order to take full advantage of the three antenna terminals on the UV-3, I needed three separate antennas, but I wanted an omnidirectional pattern with no "holes."

The solution I chose was to use three stub-fed verticals on one whip. The stub-fed vertical, or J antenna, consists of a basic half-wave radiator end fed through a quarter-wave stub. This stub serves as an impedance transformer. It transforms the high impedance of the half-wave radiator to that of the low-impedance coaxial line. Few antennas lend themselves to omnidirectional patterns and ease of matching to coaxial line as well as the stub-fed vertical.

Construction

My approach is cheap, novel and effective and uses only four basic parts except for the coaxial lines: the whip and three easily fabricated blocks. These materials are available at most hardware or hobby stores. The whip is one piece of 3/8-inch (9.5-mm) aluminum tubing 60 inches (152 mm) in length. Be sure that the piece you select is straight and free of nicks or dents.

Overall construction is shown in Figure 1. The three stub blocks are made from 3/8-inch (9.5-mm) aluminum stock. Refer to Fig. 2 and saw three blocks 3/8 × 5/8 × 1 1/8 inches (9.5 × 15.9 × 28.6mm). Drill a 3/8-inch (9.5-mm) hole as shown so that the piece will slip over the mast. Tap a no. 6-32 hole into the 3/8-inch (9.5-mm) hole just drilled for a setscrew to hold the block in place. The third hole is used to connect the braid of the coaxial cable to the mast. It is at this point where the quarter-wave stub begins and the feed line ends. For RG-58/U and similar size cable use a 13/64-inch (5.2-mm) drill and tap the hole with 1/4-20 thread. For RG-8/U, use a 25/64-inch (9.9-mm) or "X" drill and tap with 7/16-20 thread. Prepare the coaxial cables by separating the center conductors from the remainder of the cable to the lengths given in Figure 1. Cut off all but 3/8 inch (9.5 mm) of the braid and fold this back over the jacket. These sections can be threaded into the tapped holes. The blocks can then be mounted to the whip as in Figure 1.

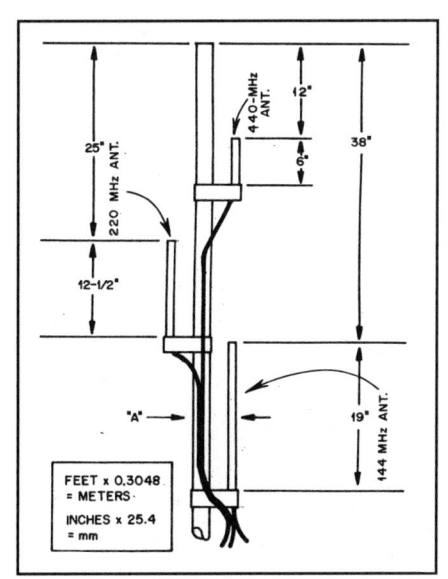

Figure 1—Construction dimensions of the three-band antenna. Cables should be routed and taped as shown.

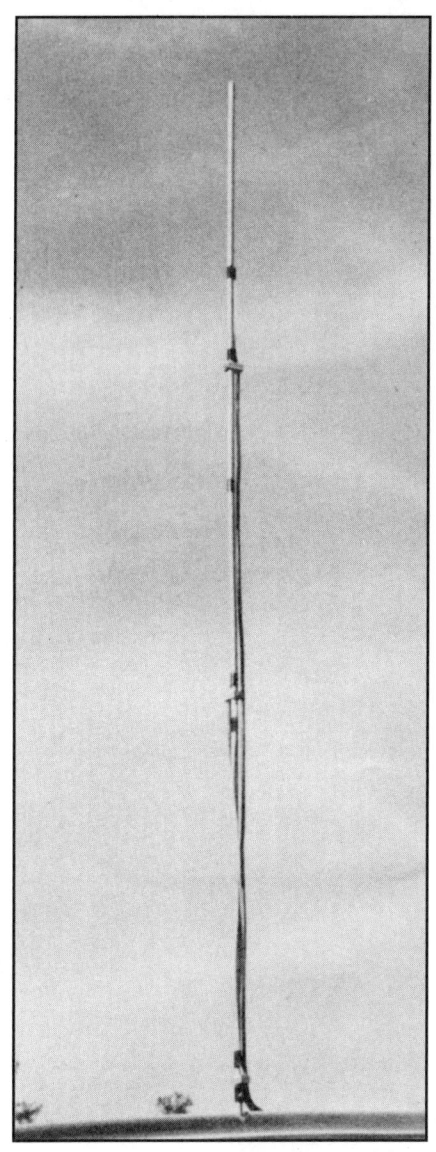

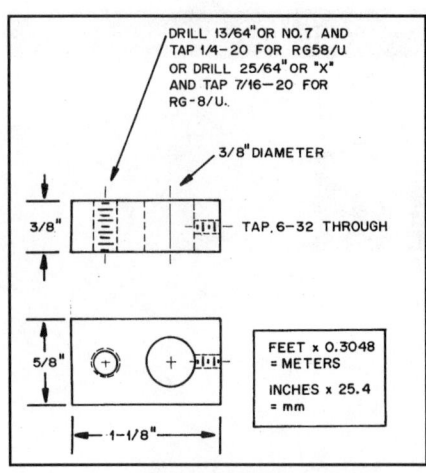

Figure 2—Detail drawing of the stub blocks used to connect and support and the quarter-wave sections.

Matching

As mentioned earlier, the quarter-wave stub is an impedance transformer. The spacing between the coaxial cable center conductor and the whip (dimension "A" in Figure 1) determines the impedance of this section and consequently the match to 50-ohm line. Using an SWR indicator, determine the optimum spacing "A." This dimension can vary greatly depending on the size of the cable and its dielectric material. Once I determined the correct spacing, I stood off the center conductor from the main support with small styrofoam blocks. Electrical tape was used to hold the quarter-wave section and styrofoam block to the main support.

The cables from the 440- and 220-MHz antennas should be routed as shown in Figure 1 on opposite sides of the main support

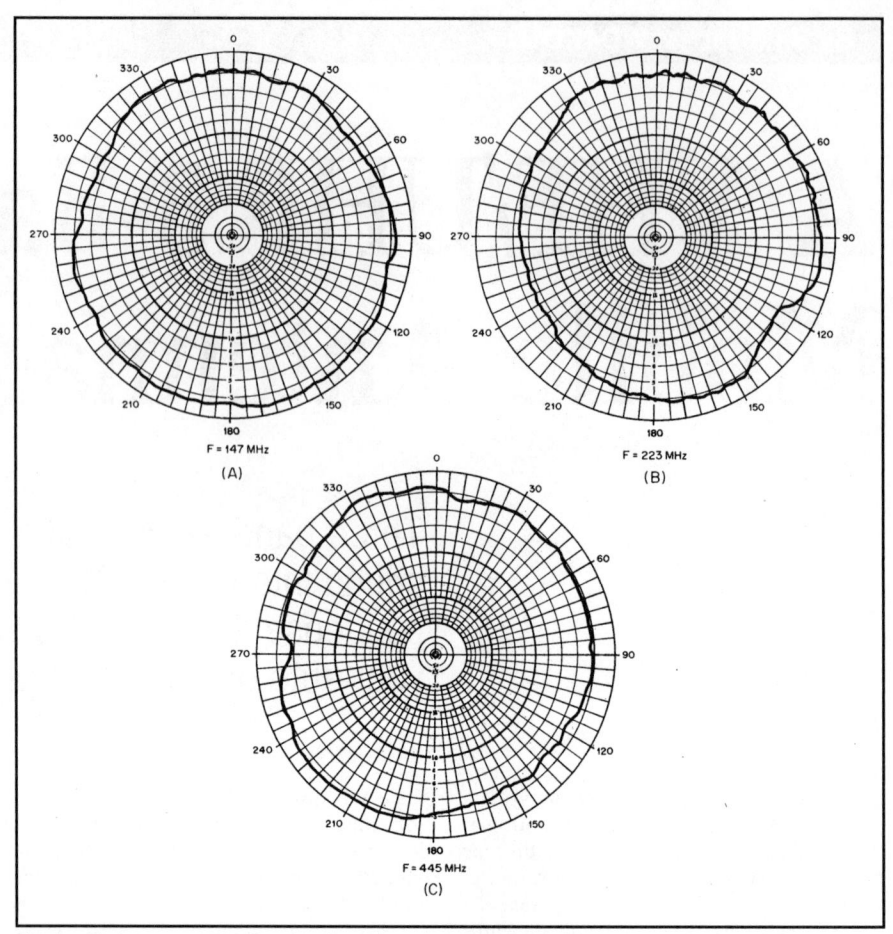

Fig 3—E-plane patterns for the three-band antenna. The patterns at A, B and C, respectively, are measued responses for 147, 223 and 445 MHz.

and away from other stubs.

The assembly is finished by taping all cables in place and coating the stub blocks with clear acrylic spray to prevent moisture from entering the cables. Although this antenna system is intended for this purpose, it should not be overlooked as a base station system. Just add 6 meters and you've got a 4-band array! E-plane patterns for the three bands are shown in Figure 3.

By Edison Fong, WD4KGD

From *QST*, March 2003

The DBJ-1: A VHF-UHF Dual-Band J-Pole

Searching for an inexpensive, high-performance dual-band base antenna for VHF and UHF? Build a simple antenna that uses a single feed line for less than $10.

Two-meter antennas are small compared to those for the lower frequency bands, and the availability of repeaters on this band greatly extends the range of lightweight low power handhelds and mobile stations. One of the most popular VHF and UHF base station antennas is the J-Pole.

The J-Pole has no ground radials and it is easy to construct using inexpensive materials. For its simplicity and small size, it offers excellent performance. Its radiation pattern is close to that of an "ideal" dipole because it is end fed; this results in virtually no disruption to the radiation pattern by the feed line.

The Conventional J-Pole

I was introduced to the twinlead version of the J-Pole in 1990 by my long-time friend, Dennis Monticelli, AE6C, and I was intrigued by its simplicity and high performance. One can scale this design to one-third size and also use it on UHF. With UHF repeaters becoming more popular in metropolitan areas, I accepted the challenge to incorporate both bands into one antenna with no degradation in performance. A common feed line would also eliminate the need for a duplexer. This article describes how to convert the traditional single band ribbon J-Pole design to dual-band operation. The antenna is enclosed in UV-resistant PVC pipe and can thus withstand the elements with only the antenna connector exposed. I have had this antenna on my roof since 1992 and it has been problem-free in the San Francisco fog.

The basic configuration of the ribbon J-Pole is shown in Figure 1. The dimensions are shown for 2 meters. This design was also discussed by KD6GLF in *QST*.[1] That antenna presented dual-band resonance, operating well at 2 meters but with a 6-7 dB deficit in the horizontal plane at UHF when compared to a dipole. This is attributable to the antenna operating at its third harmonic, with multiple out-of-phase currents.

I have tested single-band J-Pole con-

Figure 1—Basic diagram and dimensions for the original 2-meter ribbon J-Pole.

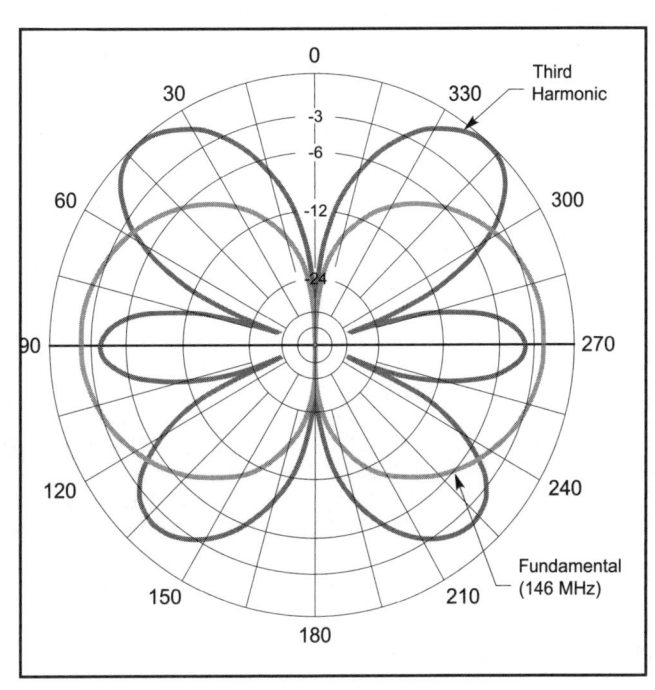

Figure 2—This EZNEC plot shows the difference in the radiation patterns between a vertical half-wave radiator operated at its fundamental (146 MHz) and third harmonic (445 MHz). At the fundamental, most of the energy is directed at right angles to the antenna—and to distant receivers.

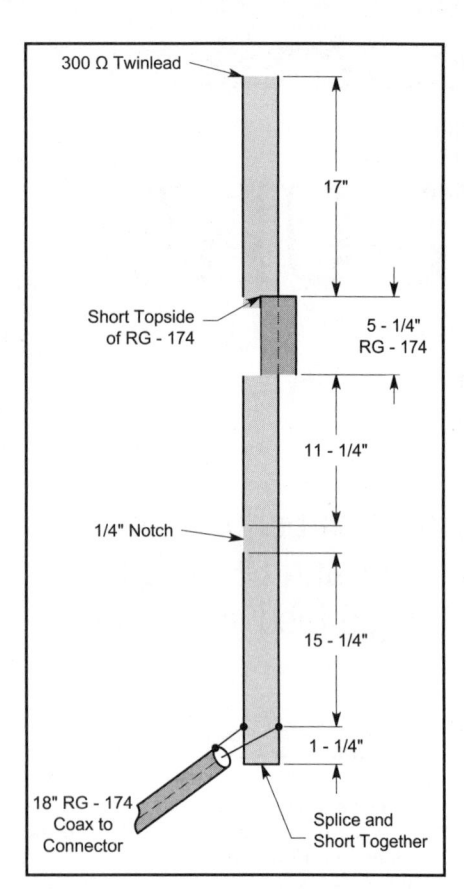

Figure 3—The 2 meter J-Pole modified for both VHF and UHF operation. These measurements are approximate (see text).

figurations constructed from copper pipe, 450 Ω ladder line, and aluminum rod. While all the designs performed well, each had shortcomings. The copper pipe J-Pole matching section would be exposed to the air, raising a durability question. The aluminum design would be faced with a similar issue in the salt air of the San Francisco Bay area. I favor the use of 300 Ω twin lead because it is easily obtainable and inexpensive. An advantage of the copper pipe design was an 8 MHz bandwidth—about twice that exhibited by the twin lead version. That was expected, since the copper pipe had a much larger diameter than the twin lead elements used in that version. My final decision was to be based on aesthetics, cost and durability...but the antenna had to be a true dual-band design.

How the J-Pole Works

The basic J-Pole antenna is a half-wave vertical radiator, much like a dipole. What separates this design from a vertical dipole is the method of feeding the half-wave element. In a conventional dipole or groundplane, the radiation pattern can be disrupted by the feed line and there is usually a tower or some other support that acts as a reflector as it is frequently parallel to the antenna. The J-Pole pattern resembles that of an ideal vertical dipole because of its minimal interaction with the feed line. The performance of this J-Pole is, theoretically at least, equal to a ¼ wave radiator over an ideal ground.

The J-Pole also matches the high impedance at the end of a ½ wave radiator to a low feed point impedance suitable for coax feed. This is done with a ¼ wave matching stub, shorted at one end and connected to the ½ wave radiator's high impedance at its other end. Between the shorted and high impedance ends there is a point that is close to 50 Ω. This is where the feed line is attached.

Creating the Dual-Band DBJ-1

So how can one add UHF to the conventional 2-meter J-Pole? First of all, a half-wave 2 meter antenna does resonate at UHF. Resonating is one thing, but working well is another. The DBJ-1 not only resonates, but also performs as a ½ wave radiator on both bands. An interesting fact to note is that ½ wave center-fed dipole-type antennas will resonate at odd harmonics (3rd, 5th, 7th, etc). This is why a 40 meter center-fed ½ wave dipole can be used on 15 meters. Similarly, a 150 MHz antenna can be used at 450 MHz. However, the performance of the antenna at the third harmonic is poor when it is used in a vertical configuration. At UHF (450 MHz) the ½ wave radiator becomes ³/₂ wavelengths long. Unfortunately, at UHF, the middle ½ wavelength is out of phase with the top and bottom segments and the resulting partial cancellation results in approximately 2 dB less gain in the horizontal plane compared to a J-Pole operating at its fundamental frequency. Maximum radiation is also directed away from the horizon. Thus, although the J-Pole can be made to work at its third harmonic, its performance is poor, often 6-8 dB below that of a groundplane. Figure 2 shows a polar plot of a vertical ½ wave radiator operating at its fundamental (146 MHz) and third harmonic (445 MHz) frequencies. Note the difference in energies of the two frequencies.

What is needed is a method to decouple the extra length of the 2 meter radiator at UHF in order to create independent ½ wavelength radiators at both VHF and UHF. The DBJ-1 accomplishes this by using a coaxial stub, as shown in the antenna drawing of Figure 3.

There is 18 inches of RG-174 transmission line connecting the bottom RF connector to the radiating element. Eighteen inches was chosen so that the bottom portion of the antenna housing can be used to mount the antenna without disturbing its electrical characteristics. [The use of RG-174 coax in this design limits the power the antenna can handle to less than 60 W at low SWR. By substituting RG-213, RG-8 or RG-58 cable, power ratings can be improved considerably. However, the length of the decoupling stub at the UHF antenna may have to be recalculated, because of the change in velocity factor (VF) of the different cable.—Ed.]

The 16½ inch matching stub of 300 Ω twin lead works like a ¼ wave stub at VHF and a ¾ wave stub at UHF with virtually no penalty, except for a slight 0.1 db loss from the extra ½ wavelength of feedline. By experimentation, the 50 Ω point was found to be 1¼ inches from the shorted end of the stub. Although the impedance at this point is slightly inductive, it is still an excellent match to 50 Ω, with an SWR of approximately 1.3:1.

Connected to the open end of the matching stub, the radiating element for UHF is 11¼ inches long. The stub and radiator are constructed of a single piece of twin lead, separated from the matching stub by a ¼ inch notch in one conductor, as shown in Figure 3. The extra wire in the twin lead radiator sections radiates along with the driven wire, creating a thick element that is shorter than its free-space equivalent. To terminate the UHF radiating section, a shorted stub, using RG-174 coaxial cable, is used. As with the input matching stub, the open end presents a high impedance and is connected to the upper end of the UHF radiating section. Note that the stub is only an open-circuit at UHF, acting as a small inductance instead, at VHF.

The RG-174 stub connects to the upper section of 300 Ω twin lead and that completes the VHF radiating element. Note that the total length of the UHF and VHF radiating elements plus the coaxial stub do not add up to a full ½ wavelength at VHF because the inductance of the coaxial stub acts to shorten the antenna slightly.

Construction Details

The dimensions given in Figure 3 should be considered a starting point for adjustment, with final tuning requiring an SWR analyzer or bridge. During the antenna's construction phase, I started at the feed point (see Figure 3) and after each section was assembled, the input SWR was checked. After the ¼ wave VHF matching section is connected to the 11¼ inch UHF ½ wave section, check the SWR at UHF. Then add the ¼ wave UHF shorted RG-174 stub. The stub will require trimming for minimum SWR at UHF. Start with the stub 10–15% long and trim the open end for lowest SWR. As a last step, add the 17 inch section of twin lead. Again, this section should be trimmed for the lowest SWR at your frequency of choice in the 2 meter band.

To weatherize the antenna, enclose it in ¾ inch schedule-200 PVC pipe with end caps. These can be obtained from your local hardware or building supply store. When sliding the antenna into the PVC tubing, I found no need to anchor the antenna once it was inside. [If larger coaxial cable is used for the stub, it is likely that the top of the antenna will require some glue or foam to hold the antenna in place because of the additional cable weight.—Ed.] The 300 Ω twin lead is sufficiently rigid so as not to bend once it is inside the pipe. Install an SO-239 connector in the bottom end cap. Once the antenna is trimmed to the desired operating frequency, glue both end caps and seal around the SO-239 connec-

Table 1
Measured Relative Performance of the Dual-Band Antenna at 146 MHz

	VHF ¼ Wave Mobile Reference	VHF Flex Antenna ("Rubber Duck")	Standard VHF J-Pole	DBJ-1 J-Pole
Received Signal Strength	−24.7 dBm	−30.5 dBm	−24.3 dBm	−23.5 dBm
Difference from Reference	0 dB	−5.8 dB	+0.4 dB	+1.2 dB

Table 2
Measured Relative Performance of the Dual-Band Antenna at 445 MHz

	UHF ¼ Wave Mobile Reference	UHF Flex Antenna ("Rubber Duck")	Standard VHF J-Pole	DBJ-1 J-Pole
Received Signal Strength	−38.8 dBm	−45.3 dBm	−45 dBm	−38.8 dBm
Difference from Reference	0 dB	−6.5 dB	−6.2 dB	0 dB

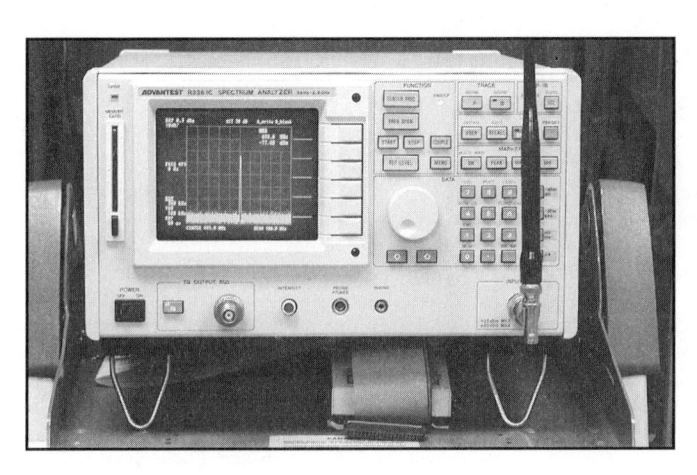

Figure 4—The Advantest R3361 spectrum analyzer used in the test.

Figure 5—The completed antenna mounted to the roof.

tor. Presto! For a few dollars, you'll have a dynamite antenna that should last for years.

The antenna should be supported only by the lower 12 inches of the housing to avoid interaction between the matching stub and any nearby metal, such as an antenna or tower. The results from the antenna are excellent considering its simplicity.

Measured Results

Brian Woodson, KE6SVX, helped me make measurements in a large parking lot, approximating a fairly good antenna range, using the Advantest R3361C spectrum analyzer shown in Figure 4.

The transmitter was a Yaesu FT-5200 located about 50 yards from the analyzer. The reference antenna consisted of mobile ¼ wave Motorola ground plane antennas mounted on an NMO connector on the top of my vehicle. The flex antenna ("rubber duck") was mounted at the end of 3 feet of coax held at the same elevation as the groundplane without radials. The J-Pole measurements were made with no groundplane and the base held at the same height as the mobile ground plane. Table 1 gives performance measurements at 146 MHz, while Table 2 gives those same measurements at 445 MHz.

As can be seen in the UHF results, the DBJ-1 outperforms the standard 2 meter J-Pole by about 6 dB (when used at UHF), a significant difference. The standard 2 meter J-Pole performance is equivalent to a flex antenna at UHF. Also note that there is no significant difference in performance at 2 meters between the DBJ-1 and a standard J-Pole. The flex antenna is about 6 dB below the ¼ wave mobile antenna at both VHF and UHF. This agrees well with the previous literature.

The completed antenna can be seen mounted to the author's roof in Figure 5.

If you do not have the equipment to construct or tune this antenna at both VHF and UHF, the completed antenna is available from the author, tuned to your desired frequency. The cost is $20. E-mail him for details.

By Brian Beezley, K6STI

From *QST*, February 1996

Another Way to Stack VHF/UHF Yagis

An unusual stacking geometry maximizes array gain and minimizes side lobes.

Yagi-Uda arrays are wonderful antennas. Invented by Shintaro Uda in collaboration with Hidetsugu Yagi in the 1920s,[1] no other antenna yields so much benefit for so little complexity in so many practical applications.[2] Yagis are easy to design, offer significant gain over simple antennas and are capable of highly directional patterns.

One of the more attractive properties of a Yagi is that it's easy to obtain higher performance by simply increasing its boom length and adding some elements. (For optimum results, however, you need to readjust all the element lengths and spacings when you resize a design, but this is easy to do with a computer using automatic optimization algorithms.[3]) There's no limit to the increase in gain and pattern quality made possible by lengthening the boom—no theoretical limit, that is! Once your Yagi becomes so long that it's mechanically infeasible to further increase the boom length, what do you do? You stack multiple Yagis, of course!

Stacking refers to the simultaneous excitation of a number of similar antennas. The term comes from the appearance of antennas arrayed above one another in the vertical plane. You can stack Yagis vertically, horizontally, or even in depth. If you space the individual antennas properly, the overall gain increases 3 dB[4] each time you double the number of antennas in a free-space array.[5] So, when your Yagi becomes unmanageably long, you simply replicate it—and make your antenna unmanageably high and wide, too!

The Stacking Trade-Off

When you stack a pair of Yagis, not only does the gain increase, but the width of the forward lobe *decreases* in the stacking plane.[6] Figure 1 shows the E-plane pattern for an 8-element, 12-foot Yagi optimized for high performance at the low end of the 2-meter band.[7] This is the azimuth pattern when the Yagi is oriented for horizontal polarization. Figure 2 shows the pattern for a pair of these antennas stacked horizontally with 150 inches between the booms. This spacing maximizes forward gain for this particular design.

Figure 2 illustrates the fundamental design trade-off for stacked arrays. Although stacking increases the array's forward gain and narrows the main lobe, it also creates a pair of large side lobes just beyond the main lobe. In this particular case, the side lobes are down only 7.5 dB from the main lobe. The side lobe level is 2 to 3 dB greater than the response of a single Yagi in the same region. This makes the stacked pair more sensitive to noise and QRM just beyond boresight (off the axis of the main lobe). Even worse, the side lobes make it much more difficult to peak the array on a received signal, especially when it's fading.

You can obtain a better pattern by reducing the stacking distance and surrendering some forward gain. Figure 3 shows the azimuth pattern for the Yagis stacked 107 inches apart. The gain has dropped 0.36 dB, but the first side lobes are now down 15 dB from the main lobe. For most applications, this is a reasonable performance compromise. If you can tolerate less gain, you can drop the side lobes further by decreasing the stacking distance even more.

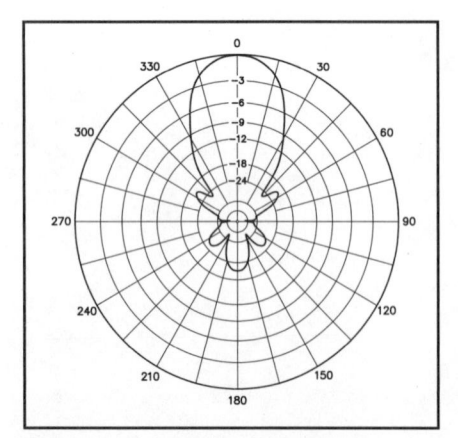

Figure 1—Free-space, E-plane pattern for an optimized, 8-element, 12-foot-boom, 2-meter Yagi. Freq: 144.2 MHz; 0 dB = 11.30 dBd.

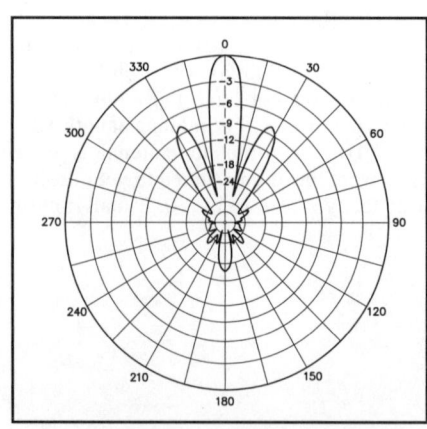

Figure 2—Free-space, E-plane pattern for a pair of 8-element Yagis stacked in the E-plane for maximum forward gain. Freq: 144.2 MHz; 0 dB = 14.38 dBd.

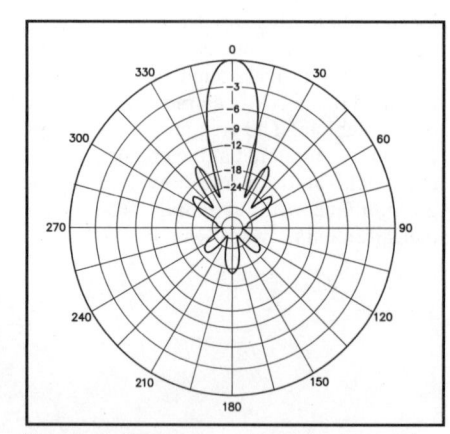

Figure 3—Free-space, E-plane pattern for a pair of 8-element Yagis stacked in the E-plane for reduced side lobes. Freq: 144.2 MHz; 0 dB = 14.02 dBd.

The Four-Stack

Many hams use a rectangular stack of four Yagis for serious weak-signal work on VHF/UHF (see Figure 4). This configuration uses two Yagis spaced horizontally with a second pair stacked directly above the first. Two vertical masts joined at their centers by a horizontal spreader support the array. This support structure is called an **H** frame. (For better stability, large arrays often use *two* horizontal spreaders displaced vertically).

The horizontal and vertical stacking distances most often are chosen to force the first side lobes in each plane to be 12 to 18 dB below the level of the main lobe. Figures 5A and 5B show the E and H-plane patterns for a four-stack using the Yagi of Figure 1, with the first side lobes down about 15 dB. The horizontal stacking distance is 107 inches and the vertical stacking distance is 99 inches.[8] This four-stack has 5.54-dB gain over a single antenna and reasonable E and H-plane patterns.

The Diamond Stack

The pattern of a stacked array can be factored into two components. The first component is called the *antenna factor* and is simply the pattern of a single array element—in this case, an individual Yagi. The second component is called the *array factor* and depends only on the stacking geometry, not on the individual antennas comprising the array. Without significant mutual-impedance interactions among individual array elements (the usual case at VHF/UHF), the pattern for the array as a whole is simply the product of the antenna factor and the array factor.

The array factor accounts for the undesirable side lobes of stacked Yagis. If you express the array geometry as a sequence of current magnitudes that correspond to the excitation of individual array elements, the array factor is simply the Fourier transform of the sequence. For the Yagi four-stack, the current sequence in either plane is 1-1. This notation indi-cates that the two Yagis are excited with equal current magnitudes. The Fourier transform of a uniform sequence like this is proportional to $(\sin x)/x$, where x is related to the angle from boresight. The first peak in this function for $x > 0$ corresponds to the undesirable first side lobes in a Yagi four-stack.

Signal-processing experts will tell you that a uniform sequence like 1-1 yields just about the worst-possible Fourier side lobes.[9] The way to lower side lobe levels is to use a tapered sequence. For example, the Fourier transform of a 1-2-1 sequence is the transform of a 1-1 sequence squared. A side lobe that's down n dB for a 1-1 array is down $2n$ dB for a 1-2-1 array of the same spacing. The theoretical trade-off is that the width of the main lobe broadens when you taper a sequence. The practical trade-off for a 1-2-1 is that an additional Yagi is required and you have to figure out how to excite it with twice the current.

Or do you?

Take a Yagi four-stack and rotate it 45° so that the array looks like a diamond instead of a square—keep the Yagis horizontal. What's the sequence of excitation currents in the horizontal plane? From left to right, we have a single Yagi, then two, and finally one. All the Yagis are in phase, so the current contribution from the center Yagis is *twice* that of the end antennas. In the horizontal plane, the current sequence is 1-2-1, and we should expect much lower side lobes! The original side lobes are still there, of course, but they're pointing harmlessly up in the air and down at the ground instead of at the horizon.

Now you can aggressively expand the spacing between Yagis to increase array gain. Figure 6 shows a diagram of the array. Figures 7A and 7B show the E and H-plane patterns for the stacking distances that maximize forward gain. The gain of the diamond array is 0.78 dB greater than that of a conventional four-stack, and the first side lobes have disappeared completely, leaving a second set of side lobes 26 dB down at 60° azimuth! The head-on, 3-D patterns in Figure 8 show the side lobe locations for conventional and diamond stacking.[10]

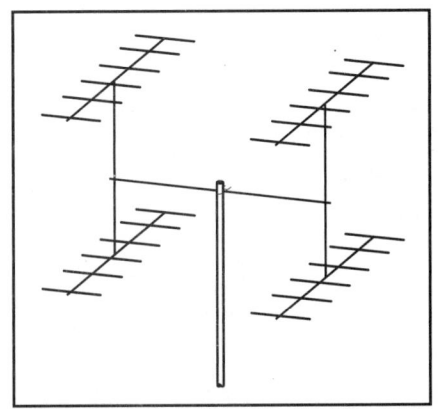

Figure 4—Stacking arrangement for a conventional Yagi four-stack.

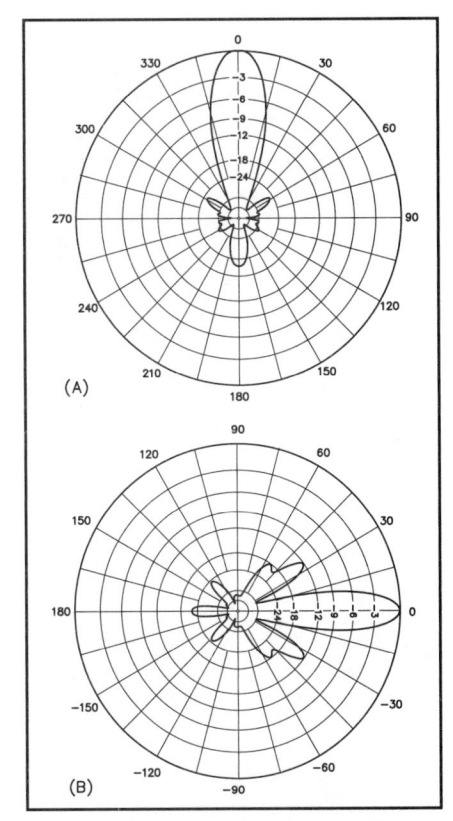

Figure 5—Free-space, E- and H-plane patterns (A and B, respectively) for a conventional Yagi four-stack. Array spacing is adjusted so that the first sidelobes in each plane are down about 15 dB from the main beam. Freq: 144.2 MHz; 0 dB = 16.84 dBd.

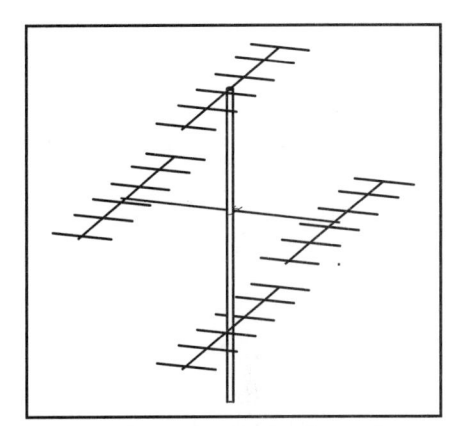

Figure 6—Arrangement of a diamond stack of four Yagis.

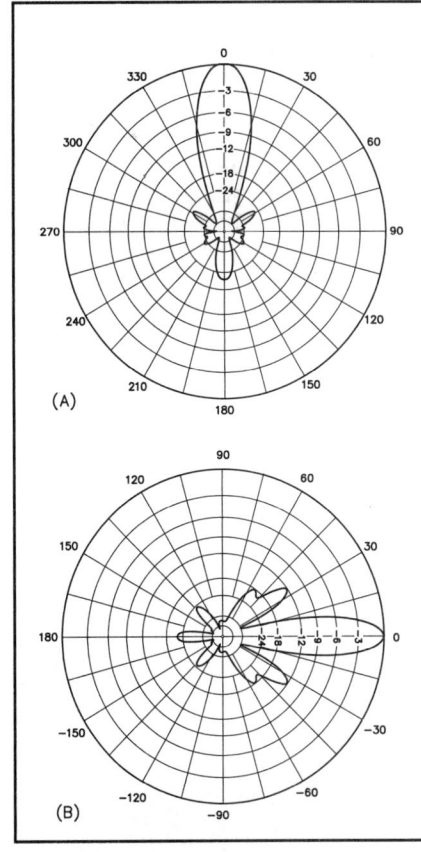

Figure 7—Free-space, E- and H-plane patterns (A and B, respectively) for a diamond stack of four Yagis. Array spacing is optimized for maximum forward gain. Freq: 144.2 MHz; 0 db = 17.60 dBd.

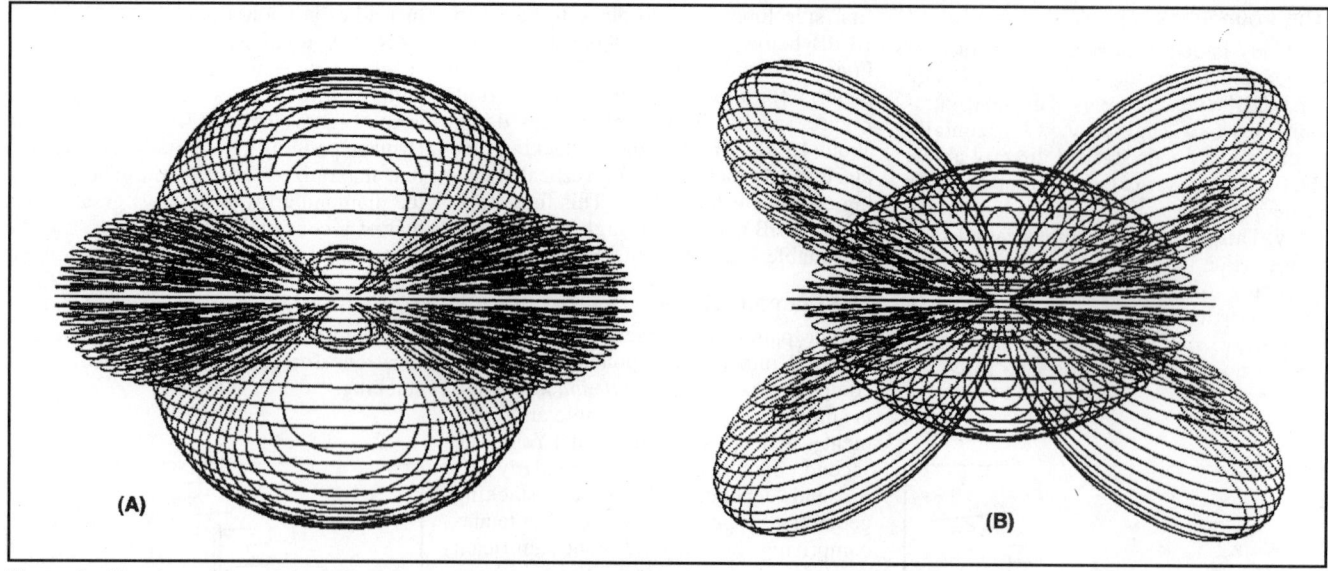

Figure 8—Head-on views of the 3-D patterns of a conventional Yagi four-stack (A) and a diamond stack (B) of the same antennas. Array spacing is the same as for Figures 5 and 7. The large, central blobs are the main beams, the small spheres are the back lobes, and the ovals are the first side lobes. Freq: 144.2 MHz; At A, peak = 16.84 dBd; at B, peak = 17.60 dBd.

The Bad News

For terrestrial use, a diamond configuration of four Yagis significantly outperforms a traditional square. But when you physically examine the array, you'll begin to appreciate the construction difficulties.

For the 12-foot Yagi used as an example, the optimum boom spacing for the diamond array is 238 inches vertically and 151 inches horizontally. To support this array, you'll need a heavy-duty, 25-foot vertical mast that extends a few feet into your tower, and a beefy, 12.6-foot horizontal spreader. In contrast, the booms of a conventional four-stack are spaced only 8.5 feet vertically and 9 feet horizontally. By contemporary VHF/UHF standards, the Yagi I've used as a building block in this article has a short boom. The stacking distances must be even greater for longer Yagis with higher gain.

Furthermore, you'll need to use an insulated section for the last few feet of horizontal spreader to avoid detuning the left and right Yagis. For the same reason, you'll need to drop their feed lines vertically for some distance rather than running them up the boom and back along the spreader. If you're using short Yagis, consider mounting them at the rear of the boom. When mounted this way, run the feed lines back along the booms and spreaders. Although unbalanced, this support method removes extraneous conductors from the array near-field, does not require insulated spreaders and keeps the feed lines tied down.

Conclusion

If you're not daunted by the construction difficulties, a diamond configuration of four Yagis offers significant performance advantages for terrestrial work when compared with a conventional square array. Above 2 meters, the diamond configuration becomes much more manageable physically. Basically, it allows you to squeeze the maximum possible performance from four Yagis. As a final bonus, the distance between Yagis is greater in an optimized diamond configuration. This reduces any residual mutual interaction between the individual antennas.

Notes

[1] John Kraus, *Antennas* (New York: McGraw-Hill, 1988), second edition, pp 481 to 482.
[2] Actually, the number-one antenna of all time for simplicity, effectiveness, and practicality has got to be the ubiquitous whip. But if we restrict our attention to directional antennas, the Yagi gets my vote.
[3] Steve Powlishen, K1FO, has developed VHF/UHF Yagi-design families that require only a simple element-length correction as elements are added. See recent editions of *The ARRL Antenna Book* or *The ARRL Handbook*. (See the *ARRL Publications Catalog* elsewhere in this issue.—*Ed.*)
[4] When stacking small Yagis in the H plane, it's possible to obtain somewhat more than 3-dB gain due to favorable mutual-impedance interactions between the antennas. For example, you can get 3.8-dB stacking gain from a pair of two-element Yagis in free space if you stack and tune them just right.
[5] This stacking-gain formula also applies at VHF/UHF when the array is many wavelengths above ground. Unfortunately, it doesn't apply for HF Yagis at typical installation heights. The elevation patterns for HF Yagis at different heights aren't very similar and don't reinforce well. For more information, see "Stacked Yagi Arrays: Fact and Fiction" by Dave Pruett, K8CC, in *NCJ*, Jul/Aug 1991, pp 18-20.
[6] The pattern in the other plane remains unchanged unless mutual impedances between the antennas alter element currents.
[7] I optimized this antenna for maximum forward gain and a good E-plane pattern with *YO 6.5*. I used *AO 6.5* to optimize stacking distance for all stacked arrays in this article. (*YO* and *AO* are available from Brian Beezley, K6STI, 3532 Linda Vista Dr, San Marcos, CA 92069; tel 619-599-4962; each program is $60. See his ad elsewhere in this issue.—*Ed.*)
[8] For equal side lobe levels, the stacking distances must differ somewhat in the two planes because Yagis have different E and H-plane patterns.
[9] The only thing worse than a 1-1-1-1 sequence, for example, is a 1-0-0-1 sequence. This amounts to disconnecting your feed line from the innermost Yagis!
[10] Dick Knadle, K2RIW, built and used diamond stacks of Yagis in the late 1970s and described them at several VHF conferences, although he never published his designs.

Brian Beezley, K6STI, can be reached at 3532 Linda Vista Dr, San Marcos, CA 92069.

By Jim Reynante, KD6GLF

From *QST*, September 1994

An Easy Dual-Band VHF/UHF Antenna

Why settle for the performance your rubber duck offers? Build this portable J-pole and boost your signal for next to nothing!

You've just opened the box that contains your new H-T and you're eager to get on the air. But the *rubber duck* antenna that came with your radio is not working well. Sometimes you can't reach the local repeater. And even when you can, your buddies tell you that your signal is noisy.

If you have 20 minutes to spare, why not build a low-cost *J-pole* antenna that's guaranteed to outperform your rubber duck? My design is a dual-band J-pole. If you own a 2-meter/70-cm H-T, this antenna will improve your signal on *both* bands.

Hams throughout the world have built and used J-pole antennas for years. My design is simple, lends itself to experimentation and alternative construction techniques, and has the following features:
- A 1.7:1 SWR or better throughout most of the 2-meter band and less than 2:1 across the 70-cm band.
- Easy set up. You can put it on the air in a matter of seconds, or store it in a space no larger than a small paperback book.
- Simple construction. The entire antenna system can be built in less than 30 minutes using TV twin lead and coaxial cable.

All of the SWR data in this article was measured at the transmitter end of the feed line. The reference impedance is 50 Ω, since most equipment is designed for this impedance.

J-Pole Antenna Theory

The J-pole is a vertically polarized antenna with two elements: the radiator and the matching stub. Although the antenna's radiator and stub are ³/₄ wavelength and ¹/₄ wavelength, respectively, it operates as an end-fed half-wave antenna. Here's how you determine the lengths of the J-pole's two elements:

$$L_{3/4} = \frac{8856 \; V}{f}$$

$$L_{1/4} = \frac{2952 \; V}{f}$$

where:
- $L_{3/4}$ = the length of the ³/₄-wavelength radiator in inches
- $L_{1/4}$ = the length of the ¹/₄-wavelength stub in inches
- V = the velocity factor of the TV twin lead
- f = the design frequency in MHz

These equations are more straightforward than they look. Just plug in the numbers and go. My design assumes that 146 MHz is the center frequency on the 2-meter band. You may, of course, substitute a center frequency of your choice. Even though the antenna is designed using a 2-meter center frequency, it also works well on 70 cm—as you'll see later.

Don't let the *velocity factor* throw you. The concept is easy to understand. Put simply, the time required for a signal to travel down a length of wire is *longer* than the time required for the same signal to travel the same distance in free space. This delay—the velocity factor—is expressed in terms of the speed of light, either as a percentage or a decimal fraction. Knowing the velocity factor is important when you're building antennas and working with

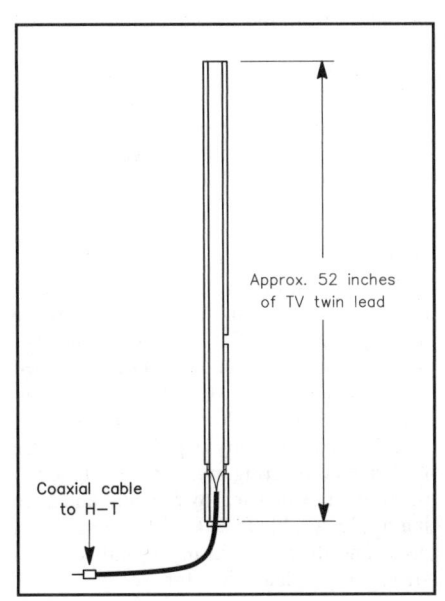

Figure 1—The J-pole antenna is approximately 52 inches long and may be hung from just about anywhere.

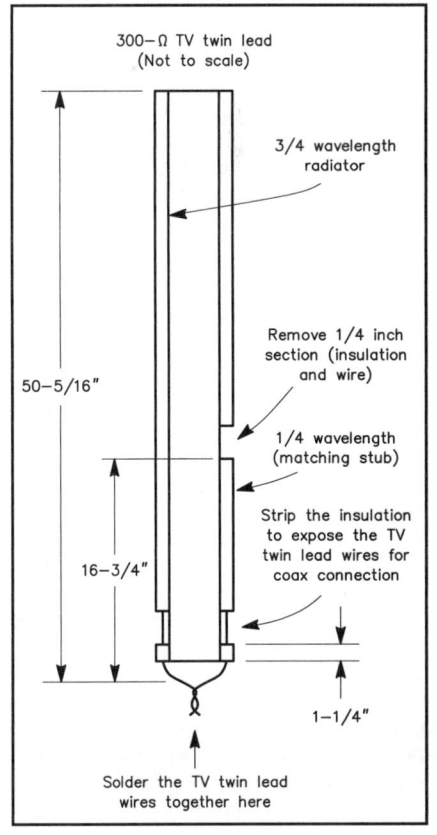

Figure 2—The basic J-pole layout. Note the areas where insulation and/or wire must be trimmed.

transmission lines. Because of the delay, 360° of a given signal wave exists in a physically *shorter* distance on a wire than in free space. This shorter distance is the *electrical* length, and that's the length we need to be concerned about.

Copper wire has a velocity factor of about 0.93, whereas TV twin lead has a velocity factor of 0.81 to 0.85 depending on who made it. If you're unsure about the twin lead you're using, just use 0.85 as its velocity factor. It's okay if it turns out to be too high. You'll be able to compensate by trimming the antenna. (It's better for the antenna to be too long than too short!) The TV twin lead I used had a velocity factor of 0.83. So, using the formulas, at 146 MHz the lengths would be approximately $50^{5}/_{16}$ inches for the $^{3}/_{4}$-wavelength radiator and $16^{3}/_{4}$ inches for the $^{1}/_{4}$-wavelength stub.

Construction

Because of the few materials needed to construct this antenna, you'll find it surprisingly easy to build. Start with approximately five feet of 300-Ω TV twin lead and about six feet of 50-Ω coaxial cable (see Figure 1) with a suitable connector (most H-Ts use a BNC connector). Use only flat 300-Ω TV twin lead, not foam core. RF can potentially short through the foam core.

Start by stripping off $^{1}/_{2}$ inch of insulation at one end of the TV twin lead (see Figure 2). Solder the two exposed wires together. This is the bottom of the antenna. Next, measure up $1^{1}/_{4}$ inches from the soldered wires and remove the insulation from the twin lead to expose $^{1}/_{8}$ to $^{1}/_{4}$ inch of wire on both sides. Be careful not to nick or break these wires. They are your connection points for the coaxial feed line.

Now you're ready to measure and cut the elements of the antenna. On one side of the twin lead, measure up $50^{5}/_{16}$ inches from the center of the exposed wire and trim off the twin lead entirely (both conductors). This side of the twin lead is the radiator of the J-pole antenna. On the opposite side of the twin lead, measure up $16^{3}/_{4}$ inches from the center of the exposed wire and carefully remove a $^{1}/_{4}$-inch section of insulation *and* wire. This is the $^{1}/_{4}$-wavelength matching stub.

Turn your attention to the coaxial cable and strip the end without the connector. Separate and expose the center conductor from the braided shield. Attach the coax to the twin lead by soldering the center conductor of the coax to the longer element of the J-pole and the shield to the shorter of the two elements. Do this at the point where you removed the twin lead insulation and exposed the wire on both sides (see Figure 3).

Apply a generous amount of weatherproof silicon sealant to the exposed coax to prevent moisture from seeping into the line. Now tape the coax to the twin lead to relieve strain on the soldered connection points. Heat shrink tubing also works well for this application.

Tuning

Hang your J-pole vertically by making a small hole at the top of the antenna and tying a length of twine or fishing line. Take care to keep the antenna away from metal objects that could detune it.

Tuning the J-pole is easy. Using a high-accuracy VHF/UHF SWR meter (borrow one if necessary), simply trim the length of the elements until you read a 1:1 SWR—or as close as you can get. Trim in *very* small increments; don't chop off an inch at a time! Remember to trim in a 3:1 ratio to maintain the $^{3}/_{4}$- to $^{1}/_{4}$-wavelength proportions. For example, if you cut $^{1}/_{8}$ inch from the $^{1}/_{4}$-wavelength stub, you must cut $^{3}/_{8}$ inches from the $^{3}/_{4}$-wavelength radiator ($^{1}/_{8} \times 3 = ^{3}/_{8}$).

I should mention that this design can cause RF coupling to the feed line. To avoid this, you can place ferrite beads on the coax at the feedpoint. An alternative is to use 3 to 5 turns of coax (1 to 2 inches in diameter) to create an RF choke at the feedpoint.

Results

Figure 4 shows my SWR measurements on 2 meters. As you can see, the antenna displayed a fairly flat SWR over most of the 2-meter band. At no point did it exceed 1.7:1. I achieved slightly higher, but useable, results on 70 cm (see Figure 5).

After hanging my J-pole from a tree limb and connecting my H-T, I switched to the frequency of a nearby repeater and gave it a try. I was able to talk with several local hams and they all said my signal sounded strong and clear. So far so good, but now came the true test. I switched to a repeater located about 17 miles north of my home, one that I couldn't use with my rubber duck antenna. I keyed the transceiver, announced my call sign, and was almost immediately greeted by a friendly voice. It worked! And not only that, it worked pretty well. The other ham said I was full-quieting into the repeater. Not bad for less than 30 minutes of work. Reception performance was also improved.

Summary

A J-pole antenna will never replace a beam or a full-size vertical mounted at 30 feet, but it offers relatively good performance for a minimum of materials, time and effort.

The applications of this antenna go beyond emergency or portable use. A permanent weatherproof enclosure can be built by mounting the J-pole inside a length of PVC tubing capped at the top. The PVC tube may then be placed at the top of a mast or similar structure. You can drill a small hole in the side of the PVC tube for the coax. Just make sure to seal it against the weather. The PVC will protect the antenna and can be painted to match the color of your house or apartment. If you live in an area where you can't put up outside antennas, hang the J-pole in your attic! If the antenna is located more than 10 feet from your radio, use a low-loss coaxial feed line such as RG-213 or equivalent.

Because of the low cost, simple construction, compact size and improved performance, there's no reason not to build several of these antennas. Keep one rolled up in your backpack when hiking, or in the glove compartment of your car!

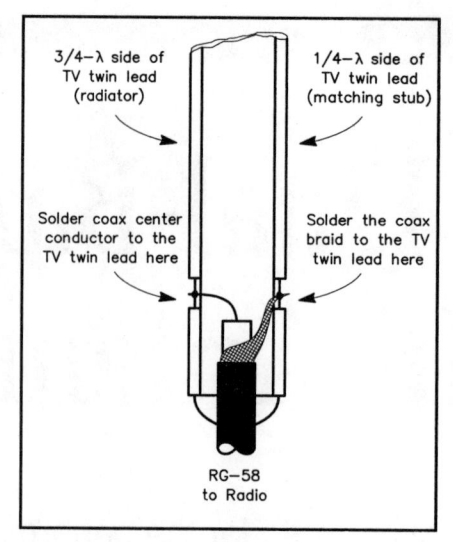

Figure 3—The coaxial feed line is connected directly at the antenna. Be careful to observe that the center conductor is soldered to the side of the TV twin lead with the longer conductor. The braid is connected to the side with the shorter conductor.

Figure 4—The SWR of the J-pole over the 2-meter band.

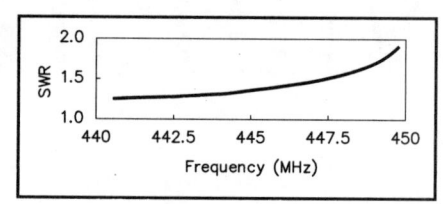

Figure 5—On the 70-cm band, the J-pole still presents a useable SWR.

AMATEUR RADIO RESOURCES

A & A Engineering
2521 West La Palma, Unit K
Anaheim, CA 92801
714-952-2114
fax 714-952-3280
e-mail w6ucm@aol.com

Absolute Value Systems
John Langner, WB2OSZ
115 Stedman St
Chelmsford, MA 01824-1823
978-256-6907
e-mail JohnL@world.std.com
web www.ultranet.com/~sstv/
  index.html

AEA
1487 Poinsettia Ave, Suite 127
Vista CA 92083
760-798-9687
800-258-7805
fax 760-798-9689
web www.AEA-wireless.com

Aero/Marine Beacon Guide
Ken Stryker
2856-G West Touhy Ave
Chicago, IL 60645

Alexander Aeroplane Co
(sold to Aircraft Spruce and
  Specialty in 1995)
Aircraft Spruce East
900 S. Pine Hill Road
Griffin, GA 30223
770-228-3901
fax 770-229-2329
web www.aircraftspruce.com

Alinco Electronics
438 Amapola Ave, #130
Torrance, CA 90501
web www.alinco.com

All Electronics Corp
PO Box 567
Van Nuys, CA 91408-0567
888-826-5432 Orders
818-904-0524 Customer Service
fax 818-781-2653
e-mail allcorp@callcorp.com
web www.allcorp.com

Allied Electronics
7410 Pebble Dr
Fort Worth, TX 76118
800-433-5700
web www.alliedelec.com

Allstar Magnetics
6205 NE 63rd St.
Vancouver, WA 98661
360-693-0213
fax 360-693-0639
web www.allstarmagnetics.com

Alpha-Delta Communications
PO Box 620
Manchester KY 40962
606-598-2029
fax 606-598-4413
web www.alphadeltacom.com

Alpha Power
Crosslink, Inc.
6185 Arapahoe Ave.
Boulder, CO 80303
303-473-9232
fax 303-473-9660
web www.alpha-amps.com

Aluma Tower Company, Inc
PO Box 2806
Vero Beach, FL 32961-2806
561-567-3423
fax 561-567-3432
e-mail atc@alumatower.com
web www.alumatower.com

AM Press/Exchange
Don Chester, K4KYV, Editor &
  Publisher
2116 Old Dover Rd
Woodlawn, TN 37191
web www.amfone.net/AMPX/
  ampx.htm

Amateur Television Quarterly
  (ATVQ)
Harlan Technologies
5931 Alma Dr
Rockford, IL 61108-2409
815-398-2683
fax 815-398-2688
e-mail atvq@hampubs.com
web www.hampubs.com/
  atvq.htm

AMECO Corporation
224 E 2nd St
Mineola, NY 11501
516-741-5030
fax 516-741-5031
e-mail sales@amecocorp.com
web www.amecocorp.com

American Design Components,
  Inc.
6 Pearl Court
Allendale, NJ 07041
800-803-5857
web www.adc-ast.com

American National Standards
  Institute (ANSI)
1819 L Street, NW
Washington, DC 20036
202-293-8020
fax 202-293-9287
web www.ansi.org

American Power Conversion
132 Fairgrounds Road
West Kingston, RI 02892
800-788-2208
web www.apcc.com

Ameritron
116 Willow Rd
Starkville, MS 39759
662-323-8211
fax 662-323-6551
e-mail amertron@ameritron.com
web www.ameritron.com

Amidon Inc.
Amidon Inductive Components
240 Briggs Ave.
Costa Mesa, CA 92626
800-898-1883
fax 714-850-1163
web www.amidon-inductive.com

AMRAD
Drawer 6148
McLean, VA 22106-6148
web www.amrad.org

AMSAT-NA (Radio Amateur
  Satellite Corp.)
850 Sligo Ave., Suite 600
Silver Spring, MD 20910-4703
301-589-6062
fax 301-608-3410
e-mail martha@amsat.org
web www.amsat.org

ANARC (Association of North
  American Radio Clubs)
Mark W. Meece, Chairman
529 Sandy Lane
Franklin, OH 45005-2065
e-mail radioscan@siscom.net
web www.anarc.org

Anchor Electronics
2040 Walsh Ave
Santa Clara, CA 95050
408-727-3693
fax 408-727-4424
web www.demoboard.com/
  anchorstore.htm

Angle Linear
PO Box 35
Lomita, CA 90717-0035
310-539-5395
fax 310-539-8738
e-mail chip@anglelinear.com
web www.anglelinear.com

AntenneX Magazine
P. O. Box 271229
Corpus Christi, TX 78427-1229
361-855-0250
888-855-9098
fax 361-855-0190
web www.antennex.com

Antique Electronic Supply
6221 South Maple Ave
PO Box 27468
Tempe AZ 85285-7468
480-820-5411
fax 480-820-4643
web www.tubesandmore.com

Antique Radio Classified
PO Box 2-V75
Carlisle, MA 01741
978-371-0512
fax 978-371-7129
web www.antiqueradio.com

Ar^2 Communication Products
Box 1242
Burlington, CT 06013
860-485-0310
fax 860-485-0311
web www.advancedreceiver.
  com

Array Solutions
350 Gloria Rd
Sunnyvale, TX 75182
972-203-2008
fax 972-203-8811
e-mail
  wx0b@arraysolutions.com
web www.arraysolutions.com

Arrow Electronics
25 Hub Dr
Melville, NY 11747
631-391-1300

ARRL—The national association
  for Amateur Radio
225 Main St
Newington, CT 06111-1494
860-594-0200
fax 860-594-0259
e-mail tis@arrl.org
web www.arrl.org

ARS Electronics
7110 De Celis Pl
PO Box 7323
Van Nuys, CA 91409
818-997-6200
800-422-4250
fax 818-997-6158

ATCI Consultants
11720 Chairman Dr, #108
Dallas, TX 75243
214-343-0600
fax 214-343-0716
e-mail atci@dallas.net
web www.dallas.net/~atci

Atlantic Surplus Sales
3730 Nautilus Ave
Brooklyn, NY 11224
718-372-0349

ATV Research, Inc
1301 Broadway
PO Box 620
Dakota City, NE 68731-0620
402-987-3771
800-392-3922 (orders)
fax 402-987-3709
e-mail sales@atvresearch.com
web www.atvresearch.com

Avantek
3175 Bowers Ave
Santa Clara, CA 95054-3292
408-727-0700

Barker and Williamson Corp
  (B&W)
603 Cidco Rd
Cocoa, FL 32926
321-639-1510
fax 321-639-2545
e-mail
  custsrvc@bwantennas.com
web www.bwantennas.com

B.G. Micro
555 N 5th St, Ste 125
Garland, TX 75040
800-276-2206
fax 972-205-9417
e-mail bgmicro@bgmicro.com
web www.bgmicro.com

Brian Beezley, K6STI
3532 Linda Vista
San Marcos, CA 92069
760-599-8662 (product support)
e-mail k6sti@n2.net

Bencher, Inc
831 N Central Ave
Wood Dale, IL 60191
630-238-1183
fax 630-238-1186
e-mail bencher@bencher.com
web www.bencher.com

Bird Electronic Corporation
30303 Aurora Rd
Cleveland, OH 44139
440-248-1200
web www.bird-electronic.com

British Amateur Television Club
Grenehurst, Pinewood Rd
High Wycombe
Bucks HP12 4DD
United Kingdom
+44-01494-528899
e-mail memsec@batc.org.uk
web www.batc.org.uk/index.htm

Buckeye Shapeform
555 Marion Rd.
Columbus, OH 43207
800-728-0776
614-445-8433
fax 614-445-8224
e-mail
  info@buckeyeshapeform.com
web
  www.buckeyeshapeform.com

Buckmaster Publishing
6196 Jefferson Highway
Mineral, VA 23117
800-282-5628 (orders)
540-894-5777
fax 540-894-9141
e-mail info@buck.com
web www.buck.com

References 115

C3i Antennas
7197 N Starcrest Dr
Warrenton, VA 20187-3579
540-349-8833
800-445-7747
web www.c3iusa.com

Caddock Electronics
1717 Chicago Ave
Riverside, CA 92507-2364
909-788-1700
fax 909-369-1151
web www.caddock.com

Calogic, LLC
237 Whitney Pl
Fremont, CA 94539
510-656-2900
fax 510-651-1076

Jim Cates, WA6GER
3241 Eastwood Rd
Sacramento, CA 95821
916-487-3580

(Cetron power tube distributor)
Richardson Electronics, Ltd.
P. O. Box 393
La Fox, IL 60147
630-208-2200
fax 630-208-2550
web www.rell.com

Circuit Specialists Inc.
220 S Country Club Dr, Bldg #2
Mesa, AZ 85210
480-464-2485
800-528-1417
fax 480-464-5824

Coilcraft
1102 Silver Lake Rd
Cary, IL 60013
847-639-6400
fax 847-639-1469
e-mail info@coilcraft.com
web www.coilcraft.com

Communication Concepts, Inc
(CCI)
508 Millstone Dr
Beavercreek, OH 45434-5840
937-426-8600
fax 937-429-3811
e-mail ccidayton@pobox.com

Communications and Power
Industries
Eimac Division
301 Industrial Way
San Carlos, CA 94070-2682
800-414-TUBE (414-8823)
fax 650-592-9988
web www.eimac.com

Communications Quarterly
(see CQ Communications)

Communications Specialists Inc
426 West Taft Ave
Orange, CA 92865-4296
714-998-3021
800-854-0547
fax 714-974-3420 or 800-850-0547
web www.com-spec.com

Condenser Products Corp
2131 Broad St
Brooksville, FL 34609
352-796-3561
888-598-0957
fax 352-799-0221

Contact East, Inc
335 Willow St
North Andover, MA 01845-5995
978-682-9844
fax 800-743-8141
e-mail sales@contacteast.com
web www.contacteast.com

Courage HANDI-HAM System
3915 Golden Valley Rd
Golden Valley, MN 55422
763-520-0512
866-426-3442 (toll free)
763-520-0245 (TTY)
fax 763-520-0577
e-mail handiham@courage.org
web www.mtn.org/handiham

CQ Communications
25 Newbridge Rd
Hicksville, NY 11801
516-681-2922 (business office)
fax 516-681-2926
e-mail cq@cq-amateur-radio.com
web www.cq-amateur-radio.com

Dave Curry Longwave Products
PO Box 1884
Burbank, CA 91507
818-846-0617
web www.fix.net/~jparker/currycom.htm

Cushcraft Corp
48 Perimeter Road
Manchester, NH 03103
603-627-7877
fax 603-627-1764
e-mail hamsales@cushcraft.com
web www.cushcraft.com

Custom Computer Services, Inc
PO Box 2452
Brookfield, WI 53008
262-797-0455
fax 262-797-0459
web www.ccsinfo.com

Jacques d'Avignon, VE3VIA
1215 Whiterock Street
Gloucester, ON K1J 1A7
Canada
613-745-6522
e-mail monitor@rac.ca

Peter W. Dahl Co., Inc.
5896 Waycross Ave.
El Paso, TX 79924
915-751-2300
fax 915-751-0768
e-mail pwdco@pwdahl.com
web www.pwdahl.com

Dallas Remote Imaging Group
4209 Meadowdale Dr.
Carrollton, TX 75010
972-898-3563
e-mail consulting@drig.com
web www.drig.com

Dan's Small Parts and Kits
Box 3634
Missoula, MT 59806-3634
406-258-2782 (voice and fax)
web www.fix.net/dans.html

Davis RF Co, Div of Davis Associates, Inc.
PO Box 730
Carlisle, MA 01741
800-328-4773 (orders)
978-369-1738 (technical info)
fax 978-369-3484
e-mail davisRFinc@aol.com
web www.davisRF.com

DC Electronics
PO Box 3203
2200 N Scottsdale Rd
Scottsdale, AZ 85271-3203
800-467-7736
fax 480-994-1707
e-mail clifton@dckits.com
web www.dckits.com/index.htm

DCI Inc
20 South Plains Road
Emerald Park, SK S4L 1B7
Canada
306-781-4451
800-563-5351
fax 306-781-2008
e-mail dci@dci.ca
web www.dci.ca

Digi-Key Corp
701 Brooks Ave S
Thief River Falls, MN 56701-0677
218-681-6674
800-344-4539 (800-DIGI-KEY)
fax 218-681-3380
web www.digikey.com

Digital Vision, Inc
270 Bridge St
Dedham, MA 02026
617-329-5400
617-329-8387 (BBS)

Dover Research
321 W 4th St
Jordan, MN 55352-1313
612-492-3913

Down East Microwave
954 Rte 519
Frenchtown, NJ 08825
908-996-3584
fax 908-996-3072
web www.downeastmicrowave.com/

East Coast Amateur Radio, Inc
314 Schenck St
N Tonawanda, NY 14120
716-695-3929
fax 716-695-0000
web www.eastcoastradio.com

EDI, Inc
1260 Karl Ct
Wauconda, IL 60084
708-487-3347

Edlie Electronics, Inc
2700 Hempstead Tpk
Levittown, NY 11756-1443
orders 800-647-4722
516-735-3330
fax 516-731-5125
e-mail Elieblinng@aol.com
web www.edlieelectronics.com

Eimac
(see Communications and Power Industries)

Electric Radio Magazine
14643 County Rd G
Cortez, CO 81321-9575
970-564-9185 (voice and fax)
e-mail er@frontier.net

Electro Sonic, Inc
100 Gordon Baker Rd
Toronto, ON M2H 3B3
Canada
416-494-1666 (outside of Canada)
416-494-1555 (sales)
fax 416-496-3030

Electronic Emporium
3621-29 E Weir Ave
Phoenix, AZ 85040
602-437-8633
fax 602-437-8835

Electronic Industries Alliance (EIA)
2500 Wilson Blvd
Arlington, VA 22201-3834
703-907-7500
web www.eia.org/

Electronic Precepts of Florida
11651 87th St
Largo, FL 34643-4917
800-367-4649
fax 727-393-1177
web www.theepgd.com/books/east/dist/365.htm

Electronic Rainbow, Inc
6227 Coffman Rd
Indianapolis, IN 46268
317-291-7262
fax 317-291-7269
web www.rainbowkits.com

Electronics Now
500-B Bi-County Blvd
Farmingdale, NY 11735
516-293-3000

Elktronics
12536 T.77
Findlay, OH 45840
419-422-8206

Elna Magnetics
234 Tinker St
PO Box 395
Woodstock, NY 12498
800-553-2870
845-679-2497
fax 845-679-7010
web http://elna-ferrite.com

Embedded Research
PO Box 92492
Rochester, NY 14692
e-mail sales@embres.com
web www.embres.com

EMI Filter Company
9075-C N 130th Ave N
Largo, FL 33773-1405
800-323-7990
727-586-7990
fax 727-586-5138
web www.emifiltercompany.com

Encomm, Inc.
1506 Capitol Ave
Plano, TX 75074
214-423-0024

Engineering Consulting
583 Candlewood St
Brea, CA 92821
714-671-2009
fax 714-255-9984

ESF Copy Service
4011 Clearview Dr
Cedar Falls, IA 50613-6111
319-266-7040

ETO, Inc
(see Alpha Power)

Fair Radio Sales Co, Inc
2395 St. Johns Road
PO Box 1105
Lima, OH 45802-1105
419-227-6573
419-223-2196
fax 419-227-1313
e-mail fairradio@fairradio.com
web www.fairradio.com/

Fair-Rite Products Corp
PO Box J, 1 Commercial Row
Wallkill, NY 12589
845-895-2055
888-324-7748
fax 845-895-2629
fax 888-337-7483
e-mail ferrites@fair-rite.com
web fair-rite.com/

Fala Electronics
PO Box 1376
Milwaukee, WI 53201-1376

FAR Circuits
18N640 Field Court
Dundee, IL 60118-9269
847-836-9148 (voice and **fax**)
email farcir@ais.net
web www.farcircuits.net/

Federal Emergency Management Agency (FEMA)
500 C Street SW
Washington, DC 20472
202-566-1600
web www.fema.gov

Gateway Electronics
8123 Page Blvd
St. Louis, MO 63130
800-669-5810
314-427-6116
fax 314-427-3147
e-mail gateway@mvp.net
web www.gatewayelex.com/

Glen Martin Engineering
13620 Old Hwy 40
Boonville, MO 65233
800-486-1223
660-882-2734
fax 660-882-7200
e-mail info@glenmartin.com
web www.glenmartin.com/

Grove Enterprises Inc
PO Box 98
Brasstown, NC 28902
800-438-8155 (orders)
828-837-9200
fax 828-837-2216
e-mail nada@grove-ent.com
web www.grove-ent.com/

HAL Communications Corp
1201 W Kenyon Rd
PO Box 365
Urbana, IL 61801-0365
217-367-7373
fax 217-367-1701
e-mail halcomm@halcomm.com
web www.halcomm.com/

Hammond Mfg Co, Inc
256 Sonwil Dr.
Cheektowaga, NY 14225-2466
716-651-0086
fax 716-651-0726
e-mail rsc@hammondmfg.com
web www.hammondmfg.com/

Hammond Mfg, Ltd
394 Edinburgh Rd, N
Guelph, ON N1H 1E5
Canada
519-822-2960
fax 519-822-0715
e-mail rsc@hammondmfg.com
web www.hammondmfg.com/

Hamtronics, Inc
65-Q Moul Rd
Hilton, NY 14468
585-392-9430
fax 585-392-9420
e-mail jv@hamtronics.com
web www.hamtronics.com/

Heathkit Educational Systems
455 Riverview Drive, Building 2
Benton Harbor, MI 49022
800-253-0570
616 925-6000
fax 616-925-2898
e-mail info@heathkit.com
web www.heathkit.com/

Henry Radio
2050 South Bundy Dr
Los Angeles, CA 90025
310-820-1234
800-877-7979 (Orders)
fax 310-826-7790
e-mail
 henryradio@earthlink.net
web www.henryradio.com/

Herbach and Rademan
 (H & R Co)
353 Crider Avenue
Moorestown, NJ 08057
800-848-8001 (Orders only)
856-802-0422
fax 856-802-0465
e-mail sales@herbach.com
web www.herbach.com/

HERD Electronics
220 S 2nd St
Dillsburg, PA 17019-9601
717-432-3248
fax 717-432-7850

Heritage Transformer Co, Inc
13483 Litchfield Rd
Eastview, KY 42732
270-862-9877
e-mail Ed-Heri-Tran@KVNet.org

Hi-Manuals
(see Surplus Sales of Nebraska)

HI-Tech Software, LLC
6600 Silacci Way
Gilroy, CA 95020
800-735-5715
fax 866 898 8329
e-mail hitech@htsoft.com
web www.htsoft.com/

Hosfelt Electronics
2700 Sunset Blvd
Steubenville, OH 43952
800-524-6464
fax 800-524-5414
e-mail order@hosfelt.com
web www.hosfelt.com/

Howard W. Sams and Company
5436 W. 78th St.
Indianapolis, IN 46268-3910
800-428-7267 (428-SAMS)
317-298-5565
fax 800.552.3910
e-mail
web www.samswebsite.com/index.html

ICOM America, Inc
2380 116th Ave NE
PO Box C-90029
Bellevue, WA 98004
425-454-8155
425-450-6088 (literature)
fax 425-454-1509
e-mail
 amateur@icomamerica.com
web www.icomamerica.com/

Idiom Press
PO Box 1025
Geyserville, CA 95441-1025
707-431-1286
E-Mail Sales@IdiomPress.com
web www.idiompress.com/

IEEE (Corporate Offices)
3 Park Avenue, 17th Floor
New York, NY 10016-5997
212-419-7900
fax 212-752-4929
web www.ieee.org/

IEEE Operations Center
445 Hoes Ln
PO Box 1331
Piscataway, NJ 08854-1331
732-981-0060
fax 732-981-1721
web www.ieee.org/

Industrial Communications Engineers (ICE)
PO Box 18495
3318 N Gale St
Indianapolis, IN 46218-0495
317-545-5412
800-423-2666
fax 317-545-9645

Industrial Safety Co
1390 Neubrecht Rd
Lima, OH 45801
877-521-9893
800-809-4805
fax 419-228-5034
fax 800-854-5498
web http://www.indlsafety.com/

International Components Corp
175 Marcus Blvd
Hauppauge, NY 11788
877-791-9477
631-952-9595 (NY)
fax 631-952-9597
e-mail oemsales@icc107.com
web www.icc107.com/

International Crystal Mfg Co
10 North Lee
PO Box 26330
Oklahoma City, OK 73126-0330
800-725-1426
405-236-3741
fax 800-322-9426
e-mail
 customerservice@icmfg.com
web www.icmfg.com/

International Radio
13620 Tyee Rd
Umpqua, OR 97486
541-459-5623
fax 541-459-5632
e-mail inrad@rosenet.net
web www.qth.com/inrad/

International Telecommunication Union (ITU)
Place des Nations
1211 Geneva 20, Switzerland
+41-22-733-7256
e-mail itumail@itu.int
web www.itu.int/

International Visual Communications Association (IVCA)
James Gaither, Jr, W4CR
PO Box 140336
Nashville, TN 37214
web http://www.mindspring.com/~sstv/

Intuitive Circuits
2275 Brinston Ave
Troy, MI 48083
248-524-1918
fax 248-524-3808
e-mail sales@icircuits.com
web www.icircuits.com

IPS Radio and Space Services
PO Box 1386
Haymarket NSW 1240
Australia
+61-2-9213-8000
fax +61-2-9213-8060
e-mail office@ips.gov.au
web www2.ips.gov.au

Jameco Electronics
1355 Shoreway Rd
Belmont, CA 94002
800-831-4242
fax 800-237-6948
e-mail info@jameco.com
web www.jameco.com/

James Millen Electronics
PO Box 4215BV
Andover, MA 01810- 0814
978-975-2711
fax 978-474-8949
e-mail info@jamesmillenco.com
web www.jamesmillenco.com

JAN Crystals
2341 Crystal Dr
PO Box 60017
Fort Myers, FL 33906-6017
800-526-9825 (JAN-XTAL)
941-936-2397
fax 941-936-3750
e-mail sales@jancrystals.com
web www.jancrystals.com/

JDR Microdevices
1850 South 10th St
San Jose, CA 95122-4108
800-538-5000 (Orders)
408-494-1400
fax 800-538-5005
e-mail sales@jdr.com
web www.jdr.com/

Jolida Inc. (Tube Factory)
10820 Guilford Rd
Suite 209
Annapolis Junction, MD 20701
301-953-2014
fax 301-498-0554
e-mail
 Jolidacorp@email.msn.com
web www.jolida.com/

K-Com
PO Box 82
Randolph OH 44265
877-242-4540
330-325-2110
fax 330-325-2525
e-mail K-ComInfo@K-ComFilters.com
web www.k-comfilters.com/

Kanga US
3521 Spring Lake Dr
Findlay, OH 45840
419-423-4604
e-mail kanga@bright.net
web www.bright.net/~kanga/kanga/

Kangaroo Tabor Software
1203 County Road 5
Farwell, TX 79325-9430
fax 806-225-4006
e-mail ku5s@wtrt.net
web www.taborsoft.com/

Kantronics
1202 East 23rd St
Lawrence, KS 66046-5099
785-842-7745
fax 785-842-2031
e-mail sales@kantronics.com
web www.kantronics.com/

Kenwood Communications Corp
2201 East Dominguez St
PO Box 22745
Long Beach, CA 90801-5745
310-639-4200 (customer support)
800-950-5005
fax 310-537-8235
web www.kenwood.net

Kepro Circuit Systems, Inc
3640 Scarlet Oak Blvd.
St. Louis, MO 63122-6606
800-325-3878
636-861-0364
fax 636-861-9109
e-mail sales@kepro.com
web www.kepro.com

Kilo-Tec
PO Box 10
Oak View, CA 93022
805-646-9645 (voice and fax)

Kirby
298 West Carmel Dr
Carmel, IN 46032
317-843-2212

Kooltronic
1700 Morse Ave
Ventura, CA 93003
805-642-8521
fax 805-658-2901
web www.kooltronic.com/

K2AW's Silicon Alley
175 Friends Ln
Westbury, NY 11590
516-334-7024
fax 516-334-7024

Lashen Electronics, Inc
21 Broadway
Denville, NJ 07834
800-552-7436
973-627-3783
fax 973-625-9501
e-mail sales@lashen.com
web www.lashen.com/

Roy Lewallen, W7EL
PO Box 6658
Beaverton, OR 97007
503-646-2885
fax 503-671-9046
e-mail w7el@eznec.com
web www.eznec.com

Lodestone Pacific
4769 E. Wesley Dr
Anaheim, CA 92807
800-694-8089
714-970-0900
fax 714-970-0800
web www.lodestonepacific.com

The Longwave Club of America
45 Wildflower Rd
Levittown, PA 19057
215-945-0543
e-mail naswa1@aol.com
web anarc.org/lwca/

M/A-COM, Inc (an AMP Company)
1011 Pawtucket Blvd
PO Box 3295
Lowell, MA 01853-3295
800-366-2266
978-442-5000
fax 978-442-5167
web www.macom.com/

M^2 Antenna Systems
7560 North Del Mar Ave
Fresno, CA 93711
559-432-8873
fax 559-432-3059
e-mail m2inc@m2inc.com
web www.m2inc.com/

MAI/Prime Parts
5736 N Michigan Rd
Indianapolis, IN 46208
317-257-6811
fax 317-257-1590
e-mail mai@iquest.net
web www.websitea.com/mai/index.html

The Manual Man
27 Walling St
Sayreville, NJ 08872-1818
732-238-8964
fax 732-238-8964

Marlin P. Jones & Associates, Inc
PO Box 12685
Lake Park, FL 33403-0685
800-652-6733 (Orders)
561-848-8236 (Tech)
fax 800-432-9937
e-mail mpja@mpja.com
web www.mpja.com/

MARS
Chief Air Force MARS
HQ AFCA/GCWM (MARS)
203 W. Losey St, Room 3065
Scott AFB, IL 62225-5222
618-229-5958
e-mail USAF.MARS@scott.af.mil
web https://public.afca.scott.af.mil/public/mars/mars1.htm

Navy-Marine Corps MARS
Chief US Navy-Marine Corps Military Affiliate Radio System (MARS)—Bldg 13
NAVCOMMU WASHINGTON
Washington, DC 20397-5161
web www.navymars.org/

Army MARS
HQ US ARMY SIGNAL COMMAND
ATTN: AFSC-OPE-MA (ARMY MARS)
Ft Huachuca, AZ 85613-5000
web www.asc.army.mil/mars/

Maxim Integrated Products
120 San Gabriel Dr
Sunnyvale, CA 94086
408-737-7600
fax 408-737-7194
web www.maxim-ic.com/

Richard Measures, AG6K
6455 LaCumbre Rd
Somis, CA 93066
805-386-3734
e-mail 2@vcnet.com
web www.vcnet/measures/

Mendelsohn Electronics Co, Inc (MECI)
340 E First St
Dayton, OH 45402
800-344-4465
937-461-3525
fax 800-344-6324
fax 937-461-3391
web www.meci.com

Metal and Cable Corp, Inc
9337 Ravenna Rd, Unit C
PO Box 117
Twinsburg, OH 44087
330-425-8455
fax 330-963-7246
web www.metal-cable.com/

MFJ Enterprises
PO Box 494
Mississippi State, MS 39762
662-323-5869
800-647-1800
fax 662-323-6551
e-mail mfj@mfjenterprises.com
web www.mfjenterprises.com/

Microchip Technology
2355 W Chandler Blvd
Chandler, AZ 85224-6199
480-792-7966
fax 480-792-4338
web www.microchip.com/

Microcraft Corp
PO Box 513Q
Thiensville, WI 53092
262-241-8144

Microwave Components of Michigan
PO Box 1697
Taylor, MI 48180
313-753-4581 (evenings)

Microwave Filter Co, Inc
6743 Kinne St
E Syracuse NY 13057
800-448-1666
315-438-4700
fax 888-411-8860
fax 315-463-1467
e-mail mfcsales@microwavefilter.com
web www.mwfilter.com/

Mini Circuits Labs
PO Box 350166
Brooklyn, NY 11235-0003
800-654-7949
718-934-4500
fax 718-332-4661
web www.minicircuits.com/

Mirage Communications
300 Industrial Park Road
Starkville, MS 39759
662-323-8287
fax 662-323-6551
web www.mirageamp.com/

Model Aviation
5151 East Memorial Dr
Muncie, IN 47302
765-288-4899
fax 765-289-4248
web http://modelaircraft.org/mag/index.htm

Morse Telegraph Club, Inc
Grand Secretary: Derek Cohn
8141 Stratford Dr
Clayton, MO 63105-3707
e-mail vibroplex@mindspring.com
web http://members.tripod.com/morse_telegraph_club/

Motorola Semiconductor Products, Inc
5005 East McDowell Rd
Phoenix, AZ 85008
512-891-2030
512-891-3773
web www.mot.com/

Mouser Electronics
1000 N Main St
Mansfield, TX 76063
800-346-6873
fax 817-804-3899
e-mail sales@mouser.com
web www.mouser.com/

Multi-Tech Industries, Inc.
64 South Main Street
P.O. Box 159
Marlboro, NJ 07746-0159
800-431-3223
fax 732-409-6695
e-mail multitech@sprynet.com
web www.multi-tech-industries.com/

National Electronics
PO Box 15417
Shawnee Mission, KS 66285
800-762-5049 (orders)
e-mail sales@national-electronics.com
web www.national-electronics.com/

National Fire Protection Association
1 Batterymarch Park
PO Box 9101
Quincy, MA 02269-9101
800-344-3555
617 770-3000
fax 617 770-0700
web www.nfpa.org/

National Semiconductor Corp
PO Box 58090
Santa Clara, CA 95052-8090
800-272-9959
408-721-5000
fax 800-432-9672
web www.national.com/

National Technical Information Service
5285 Port Royal Rd
Springfield, VA 22161
800-553-6847
703-605-6000 (sales desk)
703-487-4639 (TDD - for the hearing impaired)
fax 703-605-6900
web www.ntis.gov/

The New RTTY Journal
PO Box 236
Champaign, IL 61824-0236
217-367-7373
fax 217-367-1701
web www.rttyjournal.com/

New Sensor Corp
20 Cooper Station
New York, NY 10003
212-529-0466
800-633-5477 (orders)
fax 212-529-0486
web www.sovtek.com/

Newark Electronics
4801 N. Ravenswood Ave
Chicago, IL 60640-4496
800-463-9275
773-784-5100
fax 773-907-5339
web www.newark.com/

Noble Publishing Corp
630 Pinnacle Court
Norcross, GA 30071
770-449-6774
fax 770-448-2839
web www.noblepub.com/

NOISE/COM Co
E 64 Midland Ave
Paramus, NJ 07652
201-261-8797
fax 201-261-8339
e-mail info@noisecom.com.
web www.noisecom.com/

Northern Lights Software
P O Box 321
Canton, NY 13617
315-379-0161
fax 315-379-0161
e-mail nlsa@nlsa.com
web www.nlsa.com/

Nuts & Volts Magazine
430 Princeland Court
Corona, CA 92879
800-783-4624 (orders)
909-371-8497
fax 909-371-3052
e-mail subscribe@nutsvolts.com
web www.nutsvolts.com/

Oak Hills Research
(div of Millstone Technologies)
2460 S Moline Way
Aurora, CO 80014
800-238-8205 (orders)
303-752-3382
fax 303-745-6792
e-mail qrp@ohr.com
web www.ohr.com/

Ocean State Electronics
P. O. Box 1458
6 Industrial Drive
Westerly, RI 02891
401-596-3080
800-866-6626
fax 401-596-3590
e-mail ose@oselectronics.com
web www.oselectonics.com

Old Tech – Books & Things
498 Cross St.
Carlisle, MA 01741
978-371-2231

Osborne/McGraw-Hill
2600 10th St., 6th Floor
Berkeley, CA 94710
800-227-0900
web www.osborne.com

PacComm Packet Radio
 Systems, Inc.
7818-B Causeway Blvd.
Tampa, FL 33619-6574
800-486-7388 (Orders)
813-874-2980
fax 813-872-8696
e-mail info@paccomm.com
web www.paccomm.com

Palomar Engineers
P. O. Box 462222
Escondido, CA 92046
760-747-3343
fax 760-747-3346
e-mail Info@Palomar-
 Engineers.com
web www.Palomar-
 Engineers.com

Pasternak Enterprises
P. O. Box 16759
Irvine, CA 92623-6759
949-261-1920
fax 949-261-7451
e-mail sales@pasternak.com
web www.pasternak.com

PC Electronics
2522 Paxson Lane
Arcadia, CA 91007
626-447-4565
fax 626-447-0489
email tom@hamtv.com
web www.hamtv.com

Phillips Components
23142 Alcalde Drive, Suite A
Laguna Hills, CA 92653
949-855-4263
800-899-4263
fax 949-583-9337
e-mail
 info@phillipscomponents.net
web www.phillipscomponents.net

Phillips-Tech Electronics
P. O. Box 737
Trinidad, CA 95570
707-677-0159
fax 707-677-0934
e-mail samsphillips@cox.net
web www.phillips-tech.com

Bob Platts, G8OZP
43 Ironwalls Ln
Tutbury,
Strafordshire DE13 9NH
United Kingdom
+44-12-8381-3392
e-mail g8ozp@hotmail.com

PolyPhaser Corp.
P.O. Box 9000
Minden, NV 89423
800-325-7170
775-782-2511
fax 775-782-4476
e-mail info@polyphaser.com
web www.polyphaser.com

Popular Communications
(see CQ Communications)

Power Supply Components is
 now -
PSC Electronics
2304 Calle Del Mundo
Santa Clara, CA 95054
408-737-1333
e-mail jena@pscelex.com
web www.pscelex.com/
 index.html

Practical Wireless
Arrowsmith Court
Station Approach
Broadstone, Dorset BH18 8PW
United Kingdom
+44-1202-659910
fax +44-1202-659950
e-mail rob@pwpublishing.ltd.uk
web www.pwpublishing.ltd.uk/
 pw/index.html

Pro Distributors, Inc.
5135 A 69th St.
Lubbock, TX 79424
800-658-2027 (orders)
806-794-3692
fax 806-794-9699
web www.prodistributors.com

PSK31
web http://www.aintel.bi.ehu.es/
 psk31.html

QRP Quarterly (Subscriptions)
Mark Milburn, KQ0I
117 E. Philip St.
Des Moines, IA 50315-4114
e-mail kq0i@arrl.net
web www.arparci.org

Quantics
P. O. Box 2163
Nevada City, CA 95959-2163
e-mail dave@w9gr.com
web www.w9gr.com

R & L Electronics
1315 Maple Ave
Hamilton, OH 45011
800-221-7735
513-868-6399
fax 513-868-6574
e-mail sales@randl.com
web www.randl.com

Radio Adventures Corp.
RD #4, Box 240
Summit Drive
Franklin, PA 16323
814-437-5355
fax 814-437-5432
e-mail
 information@radioadv.com
web www.radioadv.com

Radio Amateur
 Telecommunications Society
c/o Brian Boccardi
203 Bishop Blvd
North Brunswick, NJ 08902
e-mail askrat@rats.org
web www.rats.org

Radio Bookstore and Radioware
P. O. Box 209
Rindge, NH 03461-0209
800-457-7373
603-899-6826
fax 603-899-6826
web www.radiobooks.com

RadioShack Corporation
100 Throckmorton St., Suite
 1800
Ft. Worth, TX 76102
817-415-3700
800-theshack
web www.radioshack.com

Radio Society of Great Britain
Lambda House
Cranborne Road
Potters Bar
Herts EN6 3JE
United Kingdom
+44-870-904-7373
fax +44-870-904-7374
e-mail postmaster@rsgb.org.uk
web www.rsgb.org

Radio Switch Corp.
(see Multi-Tech Industries, Inc.)

Radiokit
P. O. Box 973
Pelham, NH 03076
603-635-2235
fax 603-635-2943
e-mail km1h@juno.com

Ramsey Electronics, Inc.
793 Canning Parkway
Victor, NY 14564
585-924-4560
800-446-2295
fax 585-924-4886
e-mail sales@ramseymail.com
web www.ramseyelectronics.com

The Raymond Sarrio Company
6147 Via Serena St.
Alta Loma, CA 91701
800-413-1129 (orders)
fax 508-355-8261
e-mail sarrio@sarrio.com
web www.sarrio.com

Marius Rensen
e-mail mrensen@hffax.de
Web www.hffax.de

RF Parts Company
435 South Pacific St.
San Marcos, CA 92069
760-744-0700
800-737-2787 (orders)
fax 888-744-1943
e-mail rfp@rfparts.com
web www.rfparts.com

Rohn Industries, Inc.
6718 West Plank Road
Peoria, IL 61604
309-697-4400
fax 603-497-3244
e-mail mail@rohnnet.com
web www.rohnnet.com

S & S Associates
14102 Brown Rd.
Smithsburg, MD 21783
301-416-0661
fax 301-416-0963

Sentry Manufacturing Corp.
1201 Crystal Park
Chickasha, OK 73018-1766
405-224-6780
800-252-6780
fax 405-224-8808

SHF Microwave Parts Company
7102 West 500 South
La Porte, IN 43650
fax 219-785-4552
e-mail prutz@shfmicro.com
web www.shfmicro.com

Sky Publishing Corp.
49 Bay State Road
Cambridge, MA 02138-1200
800-253-0245
617-864-7360
fax 617-864-6117
e-mail skytel@skypub.com
web www.skypub.com

Skymoon
RR10, Box 27
Mt. Pleasant, TX 75455
web http://web.wt.net/~w5un/
 skymoon.htm

Skywave Technologies
17 Pine Knoll Road
Lexington, MA 02420
781-862-6742
e-mail skywavetec@aol.com
web http://members.aol.com/
 skywavetec/

Small Parts, Inc.
13980 NW 58th Ct
P. O. Box 4650
Miami Lakes, FL 33014-0650
800-220-4242 (Orders)
305-557-7955 (Customer
 Service)
fax 800-423-9009
e-mail parts@smallparts.com
web www.smallparts.com

Society of Wireless Pioneers Inc.
P. O. Box 86
Geyserville, CA 95441
e-mail k6dzy@direcpc.com
web www.sowp.org

Software Systems Consulting
615 South El Camino Real
San Clemente, CA 92682
949-498-5784
fax 949-498-0568

Solder-It Co.
P. O. Box 360
Chagrin Falls, OH 44022
440-247-6322
800-897-8989
fax 440-247-4630
e-mail fdoob@solder-it.com
web www.solder-it.com

Southern Electronics Supply
1909 Tulane Ave.
New Orleans, LA 70112
504-524-2343
800-447-0444
fax 504-523-1000
e-mail e-mail@southernele.com
web www.southernele.com

Sparrevohn Engineering
6911 E. 11th St.
Long Beach, CA 90815
652-799-1577
e-mail zsmrtfred@aol.com
web www.members.aol.com/
 zsmrtfred

SPEC-COM Journal
PO Box 1002
Dubuque, IA 52004-1002

Spectrum International, Inc
PO Box 1084
Concord, MA 01742
978-263-2145
fax 978-263-7008

Star Circuits
PO Box 94917
Las Vegas, NV 89193

SunLight Energy Systems
955 Manchester Ave SW
N Lawrence, OH 44666
330-832-3114
fax 330-832-4161
e-mail prosolar@sssnet.com
web www.seslogic.com/

Surplus Sales of Nebraska
1502 Jones St
Omaha, NE 68102-3112
800-244-4567 (orders)
402-346-4750
fax 402-346-2939
e-mail
 grinnell@surplussales.com
web www.surplussales.com/

Svetlana Electron Devices
8200 S. Memorial Parkway
Huntsville, AL 35802
256-882-1344
800-239-6900
fax 256-880-8077
e-mail sales@svetlana.com
web www.svetlana.com/

TAB/McGraw-Hill
Blue Ridge Summit, PA 17214-0850
800-822-8158

Tandy National Parts
(see RadioShack Corporation)

TE Systems
RLS Electronics - Distributor
1710 East Parkway
Russellville, AR 72802
888-315-7388
e-mail
rlselect@mail.cswnet.com
web www.rlselectronics.com/
ampframe.html

Teletec Corp
10101 North Blvd
Wake Forest, NC 27587
909-556-7800

Telex Communications, Inc
12000 Portland Ave South
Burnsville, MN 55337
952-884-4051
fax 952-884-0043
e-mail info@telex.com
web www.telex.com

Tempo Research Corp [see AEA]

Ten-Tec, Inc
1185 Dolly Parton Pkwy
Sevierville, TN 37862
865-453-7172
fax 865-428-4483
e-mail sales@tentec.com
web www.tentec.com/

Texas Towers
1108 Summit Ave, Suite 4
Plano, TX 75074
800-272-3467
972-422-7306 (Tech)
fax 972-881-0776
e-mail sales@texastowers.com
web www.texastowers.com/

Timewave Technology Inc
501 W. Lawson Ave.
St Paul MN 55117
651-489-5080
fax 651-489-5066
e-mail sales@timewave.com
web www.timewave.com

Toroid Corporation of Maryland
202 Northwood Dr
Salisbury, MD 21801
410-860-0300
fax 410-860-0302
e-mail sales@toroid.com
web www.toroid.com

Tri-Ex Tower Corp
7182 Rasmussen Ave
Visalia, CA 93291
800-328-2393 (orders)
209-651-7850
fax 209-651-5157

Trinity Software
James L. Tonne
7801 Rice Dr
Rowlett, TX 75088
972-475-7132

Tucson Amateur Packet Radio
8987-309 E Tanque Verde Rd, #337
Tucson, AZ 85749-9399
972-671-8277
fax 971-671-8716
e-mail tapr@tapr.org
web www.tapr.org

TX RX Systems, Inc
8625 Industrial Pky
Angola, NY 14006
716-549-4700
fax 716-549-4772
e-mail sales@txrx.com
web www.txrx.com

Typetronics
PO Box 8873
Fort Lauderdale, FL 33310-8873
954-583-1340
fax 954-583-0777

Unified Microsystems
PO Box 133-W
Slinger, WI 53086
262-644-9036
fax 262-644-9036
e-mail w9xt@qth.com
web www.qth.com/w9xt

United Nations Bookshop
UN General Assembly Building, Room 32B
New York, NY 10017
e-mail bookshop@un.org
web www.un.org/Pubs/
bookshop/bookshop.htm

Universal Manufacturing Co
43900 Groesbeck Hwy
Clinton Township, MI 48036
586-463-2560
fax 586-463-2964

US Electronics (Port Jefferson, NY)
Acquired by Communication Dynamics Inc.
325 Laudermilch Rd.
Hershey, PA 17033
717-312-1159

US Government Printing Office - Bookstore
202-512-1800
866-512-1800 (toll free)
fax 202-512-2250
e-mail orders@gpo.gov
web www.bookstore.gpo.gov/

US Plastic Corp
1390 Neubrecht Rd
Lima, OH 45801-3196
800-809-4217
fax 800-854-5498
e-mail usp@usplastic.com
web www.usplastic.com

US Tower Corp
1220 Marcin St
Visalia, CA 93291-9288
559-733-2438
fax 559-733-7194
e-mail sales@ustower.com
web www.ustower.com

VHF PAK
Bob Mobile, K1SIX
33 Kimball Hill Road
Hillsboro, NH 03244

VK3UM EME Planner
Doug McArthur
Tikaluna
26 Old Murrindindi Road
Glenburn, VIC 3717
Australia
Download: www.qsl.net/
sm2cew/download.htm

W&W Manufacturing Co
800 South Broadway
Hicksville, NY 11801-5017
800-221-0732
516-942-0011
fax 516-942-1944
web www.ww-manufacturing.com/

The W5YI Group
PO Box 565101
Dallas, TX 75356
800-669-9594 (orders)
817-274-0400
fax 817-548-9594
e-mail w5yi@w5yi.org
web www.w5yi.org/

W6EL Software
11058 Queensland St
Los Angeles, CA 90034-3029
310-473-7322
e-mail ad363@lafn.org

W7FG Vintage Manuals
402731 West 2155 Drive
Bartlesville, OK 74006
800-807-6146 (orders)
918-333-3754
fax 918-774-9180
e-mail w7fg@w7fg.com
web www.w7fg.com/

Ed Wetherhold, W3NQN
1426 Catlyn Pl
Annapolis, MD 21401-4208
410-268-0916
fax 410-268-4779

Wilderness Radio
PO Box 734
Los Altos, CA 94023-0734
650-494-3806
e-mail qrpbob@datatamers.com
web www.fix.net/jparker/
wild.html

Winegard
3000 Kirkwood St
Burlington, IA 52601-1007
800-288-8094
319-754-0600
fax 319-754-0787
web www.winegard.com/

The Wireman Inc
261 Pittman Rd
Landrum, SC 29356-9544
800-727-WIRE (800-727-9473)
Orders only
864-895-4195 Technical
fax 864-895-5811
e-mail info@thewireman.com
web www.thewireman.com/

Worldradio
2120 28th St
Sacramento, CA 95818
916-457-3655
877 472-8643 (subscriptions)
e-mail editor@wr6wr.com
web www.wr6wr.com/

Wyman Research, Inc
8339 South, 850 West
Waldron, IN 46182-9644
765-525-6452
e-mail wyman@svs.net
web www.svs.net/wyman

Yaesu U.S.A.
Vertex Standard
10900 Walker St
Cypress, CA 90630
714-827-7600
fax 714-827-8100
e-mail
amateursales@vxstdusa.com
web www.vxstdusa.com/

E.H. Yost and Company
Mr. NiCd's Batteries America
2211-D Parview Rd
Middleton, WI 53562
800-308-4805 Orders only
fax 608-831-1082
e-mail ehyost@chorus.net
web www.batteriesamerica.com/

Zero Surge Inc
889 State Rte 12
Frenchtown NJ 08825
800-996-6696
908-996-7700
fax 908-996-7773
e-mail info@zerosurge.com
web www.zerosurge.com/

73 Amateur Radio Today
70 Route 202 N
Peterborough, NH 03458-1107
800-274-7373 (subscriptions)
603-924-0058
fax 603-924-8613

Standard Sizes of Aluminum Tubing
6061-T6 (61S-T6) Round Aluminum Tube in 12-ft Lengths

OD (in.)	Wall Thickness (in.)	stubs ga	ID (in.)	Approx Weight (lb) per ft	per length
3/16	0.035	no. 20	0.117	0.019	0.228
	0.049	no. 18	0.089	0.025	0.330
1/4	0.035	no. 20	0.180	0.027	0.324
	0.049	no. 18	0.152	0.036	0.432
	0.058	no. 17	0.134	0.041	0.492
5/16	0.035	no. 20	0.242	0.036	0.432
	0.049	no. 18	0.214	0.047	0.564
	0.058	no. 17	0.196	0.055	0.660
3/8	0.035	no. 20	0.305	0.043	0.516
	0.049	no. 18	0.277	0.060	0.720
	0.058	no. 17	0.259	0.068	0.816
	0.065	no. 16	0.245	0.074	0.888
7/16	0.035	no. 20	0.367	0.051	0.612
	0.049	no. 18	0.339	0.070	0.840
	0.065	no. 16	0.307	0.089	1.068
1/2	0.028	no. 22	0.444	0.049	0.588
	0.035	no. 20	0.430	0.059	0.708
	0.049	no. 18	0.402	0.082	0.948
	0.058	no. 17	0.384	0.095	1.040
	0.065	no. 16	0.370	0.107	1.284
5/8	0.028	no. 22	0.569	0.061	0.732
	0.035	no. 20	0.555	0.075	0.900
	0.049	no. 18	0.527	0.106	1.272
	0.058	no. 17	0.509	0.121	1.452
	0.065	no. 16	0.495	0.137	1.644
3/4	0.035	no. 20	0.680	0.091	1.092
	0.049	no. 18	0.652	0.125	1.500
	0.058	no. 17	0.634	0.148	1.776
	0.065	no. 16	0.620	0.160	1.920
	0.083	no. 14	0.584	0.204	2.448
7/8	0.035	no. 20	0.805	0.108	1.308
	0.049	no. 18	0.777	0.151	1.810
	0.058	no. 17	0.759	0.175	2.100
	0.065	no. 16	0.745	0.199	2.399
1	0.035	no. 20	0.930	0.123	1.467
	0.049	no. 18	0.902	0.170	2.040
	0.058	no. 17	0.884	0.202	2.424
	0.065	no. 16	0.870	0.220	2.640
	0.083	no. 14	0.834	0.281	3.372
1 1/8	0.035	no. 20	1.055	0.139	1.668
	0.058	no. 17	1.009	0.228	2.736
1 1/4	0.035	no. 20	1.180	0.155	1.860
	0.049	no. 18	1.152	0.210	2.520
	0.058	no. 17	1.134	0.256	3.072
	0.065	no. 16	1.120	0.284	3.408
	0.083	no. 14	1.084	0.357	4.284
1 3/8	0.035	no. 20	1.305	0.173	2.076
	0.058	no. 17	1.259	0.282	3.384
1 1/2	0.035	no. 20	1.430	0.180	2.160
	0.049	no. 18	1.402	0.260	3.120
	0.058	no. 17	1.384	0.309	3.708
	0.065	no. 16	1.370	0.344	4.128
	0.083	no. 14	1.334	0.434	5.208
	*0.125	1/8"	1.250	0.630	7.416
	*0.250	1/4"	1.000	1.150	14.823
1 5/8	0.035	no. 20	1.555	0.206	2.472
	0.058	no. 17	1.509	0.336	4.032
1 3/4	0.058	no. 17	1.634	0.363	4.356
	0.083	no. 14	1.584	0.510	6.120
1 7/8	0.508	no. 17	1.759	0.389	4.668
2	0.049	no. 18	1.902	0.350	4.200
	0.065	no. 16	1.870	0.450	5.400
	0.083	no. 14	1.834	0.590	7.080
	*0.125	1/8"	1.750	0.870	9.960
	*0.250	1/4"	1.500	1.620	19.920
2 1/4	0.049	no. 18	2.152	0.398	4.776
	0.065	no. 16	2.120	0.520	6.240
	0.083	no. 14	2.084	0.660	7.920
2 1/2	0.065	no. 16	2.370	0.587	7.044
	0.083	no. 14	2.334	0.740	8.880
	*0.125	1/8"	2.250	1.100	12.720
	*0.250	1/4"	2.000	2.080	25.440
3	0.065	no. 16	2.870	0.710	8.520
	*0.125	1/8"	2.700	1.330	15.600
	*0.250	1/4"	2.500	2.540	31.200

*These sizes are extruded; all other sizes are drawn tubes.

Shown here are standard sizes of aluminum tubing that are stocked by most aluminum suppliers or distributors in the United States and Canada.

Nominal Characteristics of Commonly Used Transmission Lines

RG or Type	Part Number	Nom. Z_0 Ω	VF %	Cap. pF/ft	Cent. Cond. AWG	Diel. Type	Shield Type	Jacket Matl	OD inches	Max V (RMS)	1 MHz	10	100	1000
RG-6	Belden 1694A	75	82	16.2	#18 Solid BC	FPE	FC	P1	0.275	600	0.2	0.7	1.8	5.9
RG-6	Belden 8215	75	66	20.5	#21 Solid CCS	PE	D	PE	0.332	2700	0.4	0.8	2.7	9.8
RG-8	Belden 7810A	50	86	23.0	#10 Solid BC	FPE	FC	PE	0.405	600	0.1	0.4	1.2	4.0
RG-8	TMS LMR400	50	85	23.9	#10 Solid CCA	FPE	FC	PE	0.405	600	0.1	0.4	1.3	4.1
RG-8	Belden 9913	50	84	24.6	#10 Solid BC	ASPE	FC	P1	0.405	600	0.1	0.4	1.3	4.5
RG-8	CXP 9913FX	50	84	24.0	#10 Flex BC	ASPE	FC	P2N	0.405	600	0.1	0.4	1.3	4.5
RG-8	Belden 9913F7	50	83	24.6	#11 Flex BC	FPE	FC	P1	0.405	600	0.2	0.6	1.5	4.8
RG-8	Belden 9914	50	82	24.8	#10 Solid BC	FPE	FC	P1	0.405	600	0.2	0.5	1.5	4.8
RG-8	TMS LMR400UF	50	85	23.9	#10 Flex BC	FPE	FC	PE	0.405	600	0.1	0.4	1.4	4.9
RG-8	DRF-BF	50	84	24.5	#9.5 Flex BC	FPE	FC	PE	0.405	600	0.1	0.5	1.6	5.2
RG-8	WM CQ106	50	84	24.5	#9.5 Flex BC	FPE	FC	P2N	0.405	600	0.2	0.6	1.8	5.3
RG-8	CXP18214	50	78	26.0	#13 Flex BC	FPE	S	P1	0.405	600	0.1	0.5	1.8	7.1
RG-8	Belden 8237	52	66	29.5	#13 Flex BC	PE	S	P1	0.405	3700	0.2	0.6	1.9	7.4
RG-8X	Belden 7808A	50	86	23.5	#15 Solid BC	FPE	FC	PE	0.240	600	0.2	0.7	2.3	7.4
RG-8X	TMS LMR240	50	84	24.2	#15 Solid BC	FPE	FC	PE	0.242	300	0.2	0.8	2.5	8.0
RG-8X	WM CQ118	50	82	25.0	#16 Flex BC	FPE	FC	P2N	0.242	300	0.3	0.9	2.8	8.4
RG-8X	TMS LMR240UF	50	84	24.2	#15 Flex BC	FPE	FC	PE	0.242	300	0.2	0.8	2.8	9.6
RG-8X	Belden 9258	50	82	24.8	#16 Flex BC	FPE	S	P1	0.242	600	0.3	0.9	3.1	11.2
RG-9	Belden 8242	51	66	30.0	#13 Flex SPC	PE	SCBC	P2N	0.420	5000	0.2	0.6	2.1	8.2
RG-11	Belden 8213	75	84	16.1	#14 Solid BC	FPE	S	PE	0.405	600	0.2	0.4	1.3	5.2
RG-11	Belden 8238	75	66	20.5	#18 Flex TC	PE	S	P1	0.405	600	0.2	0.7	2.0	7.1
RG-58	Belden 7807A	50	85	23.7	#18 Solid BC	FPE	FC	PE	0.195	300	0.3	1.0	3.0	9.7
RG-58	TMS LMR200	50	83	24.5	#17 Solid BC	FPE	FC	PE	0.195	300	0.3	1.0	3.2	10.5
RG-58	WM CQ124	52	66	28.5	#20 Solid BC	PE	S	PE	0.195	1400	0.4	1.3	4.3	14.3
RG-58	Belden 8240	52	66	28.5	#20 Solid BC	PE	S	P1	0.193	1900	0.3	1.1	3.8	14.5
RG-58A	Belden 8219	53	73	26.5	#20 Flex TC	FPE	S	P1	0.195	300	0.4	1.3	4.5	18.1
RG-58C	Belden 8262	50	66	30.8	#20 Flex TC	PE	S	P2N	0.195	1400	0.4	1.4	4.9	21.5
RG-58A	Belden 8259	50	66	30.8	#20 Flex TC	PE	S	P1	0.192	1900	0.4	1.5	5.4	22.8
RG-59	Belden 1426A	75	83	16.3	#20 Solid BC	FPE	S	P1	0.242	300	0.3	0.9	2.6	8.5
RG-59	CXP 0815	75	82	16.2	#20 Solid BC	FPE	S	P1	0.232	300	0.5	0.9	2.2	9.1
RG-59	Belden 8212	75	78	17.3	#20 Solid CCS	FPE	S	P1	0.242	300	0.6	1.0	3.0	10.9
RG-59	Belden 8241	75	66	20.4	#23 Solid CCS	PE	S	P1	0.242	1700	0.6	1.1	3.4	12.0
RG-62A	Belden 9269	93	84	13.5	#22 Solid CCS	ASPES		P1	0.240	750	0.3	0.9	2.7	8.7
RG-62B	Belden 8255	93	84	13.5	#24 Flex CCS	ASPES		P2N	0.242	750	0.3	0.9	2.9	11.0
RG-63B	Belden 9857	125	84	9.7	#22 Solid CCS	ASPES		P2N	0.405	750	0.2	0.5	1.5	5.8
RG-142	CXP 183242	50	69.5	29.4	#19 Solid SCCS	TFE	D	FEP	0.195	1900	0.3	1.1	3.8	12.8
RG-142B	Belden 83242	50	69.5	29.0	#19 Solid SCCS	TFE	D	TFE	0.195	1400	0.3	1.1	3.9	13.5
RG-174	Belden 7805R	50	73.5	26.2	#25 Solid BC	FPE	FC	P1	0.110	300	0.6	2.0	6.5	21.3
RG-174	Belden 8216	50	66	30.8	#26 Flex CCS	PE	S	P1	0.110	1100	1.9	3.3	8.4	34.0
RG-213	Belden 8267	50	66	30.8	#13 Flex BC	PE	S	P2N	0.405	3700	0.2	0.6	1.9	8.0
RG-214	Belden 8268	50	66	30.8	#13 Flex SPC	PE	D	P2N	0.425	3700	0.2	0.6	1.9	8.0
RG-216	Belden 9850	75	66	20.5	#18 Flex TC	PE	D	P2N	0.425	3700	0.2	0.7	2.0	7.1
RG-217	WM CQ217F	50	66	30.8	#10 Flex BC	PE	D	PE	0.545	7000	0.1	0.4	1.4	5.2
RG-217	M17/78-RG217	50	66	30.8	#10 Solid BC	PE	D	P2N	0.545	7000	0.1	0.4	1.4	5.2
RG-218	M17/79-RG218	50	66	29.5	#4.5 Solid BC	PE	S	P2N	0.870	11000	0.1	0.2	0.8	3.4
RG-223	Belden 9273	50	66	30.8	#19 Solid SPC	PE	D	P2N	0.212	1400	0.4	1.2	4.1	14.5
RG-303	Belden 84303	50	69.5	29.0	#18 Solid SCCS	TFE	S	TFE	0.170	1400	0.3	1.1	3.9	13.5
RG-316	CXP TJ1316	50	69.5	29.4	#26 Flex BC	TFE	S	FEP	0.098	1200	1.2	2.7	8.0	26.1
RG-316	Belden 84316	50	69.5	29.0	#26 Flex SCCS	TFE	S	FEP	0.096	900	1.2	2.7	8.3	29.0
RG-393	M17/127-RG393	50	69.5	29.4	#12 Flex SPC	TFE	D	FEP	0.390	5000	0.2	0.5	1.7	6.1
RG-400	M17/128-RG400	50	69.5	29.4	#20 Flex SPC	TFE	D	FEP	0.195	1400	0.4	1.1	3.9	13.2
LMR500	TMS LMR500UF	50	85	23.9	#7 Flex BC	FPE	FC	PE	0.500	2500	0.1	0.4	1.2	4.0
LMR500	TMS LMR500	50	85	23.9	#7 Solid CCA	FPE	FC	PE	0.500	2500	0.1	0.3	0.9	3.3
LMR600	TMS LMR600	50	86	23.4	#5.5 Solid CCA	FPE	FC	PE	0.590	4000	0.1	0.2	0.8	2.7
LMR600	TMS LMR600UF	50	86	23.4	#5.5 Flex BC	FPE	FC	PE	0.590	4000	0.1	0.2	0.8	2.7

RG or Type	Part Number	Nom. Z_0 Ω	VF %	Cap. pF/ft	Cent. Cond. AWG	Diel. Type	Shield Type	Jacket Matl	OD inches	Max V (RMS)	Matched Loss (dB/100) 1 MHz	10	100	1000
LMR1200	TMS LMR1200	50	88	23.1	#0 Copper Tube	FPE	FC	PE	1.200	4500	0.04	0.1	0.4	1.3
Hardline														
1/2"	CATV Hardline	50	81	25.0	#5.5 BC	FPE	SM	none	0.500	2500	0.05	0.2	0.8	3.2
1/2"	CATV Hardline	75	81	16.7	#11.5 BC	FPE	SM	none	0.500	2500	0.1	0.2	0.8	3.2
7/8"	CATV Hardline	50	81	25.0	#1 BC	FPE	SM	none	0.875	4000	0.03	0.1	0.6	2.9
7/8"	CATV Hardline	75	81	16.7	#5.5 BC	FPE	SM	none	0.875	4000	0.03	0.1	0.6	2.9
LDF4-50A Heliax –1/2"		50	88	25.9	#5 Solid BC	FPE	CC	PE	0.630	1400	0.05	0.2	0.6	2.4
LDF5-50A Heliax –7/8"		50	88	25.9	0.355" BC	FPE	CC	PE	1.090	2100	0.03	0.10	0.4	1.3
LDF6-50A Heliax – 1 1/4"		50	88	25.9	0.516" BC	FPE	CC	PE	1.550	3200	0.02	0.08	0.3	1.1
Parallel Lines														
TV Twinlead (Belden 9085)		300	80	4.5	#22 Flex CCS	PE	none	P1	0.400	**	0.1	0.3	1.4	5.9
Twinlead (Belden 8225)		300	80	4.4	#20 Flex BC	PE	none	P1	0.400	8000	0.1	0.2	1.1	4.8
Generic Window Line		405	91	2.5	#18 Solid CCS	PE	none	P1	1.000	10000	0.02	0.08	0.3	1.1
WM CQ 554		420	91	2.7	#14 Flex CCS	PE	none	P1	1.000	10000	0.02	0.08	0.3	1.1
WM CQ 552		440	91	2.5	#16 Flex CCS	PE	none	P1	1.000	10000	0.02	0.08	0.3	1.1
WM CQ 553		450	91	2.5	#18 Flex CCS	PE	none	P1	1.000	10000	0.02	0.08	0.3	1.1
WM CQ 551		450	91	2.5	#18 Solid CCS	PE	none	P1	1.000	10000	0.02	0.08	0.3	1.1
Open-Wire Line		600	92	1.1	#12 BC	none	none	none	**	12000	0.02	0.06	0.2	0.7

Approximate Power Handling Capability (1:1 SWR, 40°C Ambient):

	1.8 MHz	7	14	30	50	150	220	450	1 GHz
RG-58 Style	1350	700	500	350	250	150	120	100	50
RG-59 Style	2300	1100	800	550	400	250	200	130	90
RG-8X Style	1830	840	560	360	270	145	115	80	50
RG-8/213 Style	5900	3000	2000	1500	1000	600	500	350	250
RG-217 Style	20000	9200	6100	3900	2900	1500	1200	800	500
LDF4-50A	38000	18000	13000	8200	6200	3400	2800	1900	1200
LDF5-50A	67000	32000	22000	14000	11000	5900	4800	3200	2100
LMR500	18000	9200	6500	4400	3400	1900	1600	1100	700
LMR1200	52000	26000	19000	13000	10000	5500	4500	3000	2000

Legend:

**	Not Available or varies	FC	Foil + Tinned Copper Braid	SC	Silver Coated Braid
ASPE	Air Spaced Polyethylene	FEP	Teflon ® Type IX	SCBC	SC and BC
BC	Bare Copper	Flex	Flexible Stranded Wire	SCCS	Silver Plated Copper Coated Steel
CC	Corrugated Copper	FPE	Foamed Polyethylene	SM	Smooth Aluminum
CCA	Copper Cover Aluminum	Heliax	Andrew Corp Heliax	SPC	Silver Plated Copper
CCS	Copper Covered Steel	N	Non-Contaminating	TC	Tinned Copper
CXP	Cable X-Perts, Inc.	P1	PVC, Class 1	TFE	Teflon®
D	Double Copper Braids	P2	PVC, Class 2	TMS	Times Microwave Systems
DRF	Davis RF	PE	Polyethylene	UF	Ultra Flex
		S	Single Braided Shield	WM	Wireman

This and a World Grid Locator Map are available from ARRL.

Antenna Products Suppliers

ARRL maintains an online resource to help locate suppliers for all your Amateur Radio needs. Point your Web browser to **www.arrl.org/tis/tisfind.html**. This page will allow you to search the ARRL TIS (Technical Information Service) database of more than 2000 suppliers who can provide goods and services of interest to radio amateurs. These include manufacturers, dealers, publications, clubs, museums, and so on. Or you can browse a list of categories of products and services. You may also download **tisfinst.exe** to use this database off-line.

You may search for company name, product, in fact any word contained in the record. Most common searches are for addresses, telephone numbers and e-mail addresses. You can enter a company name (full or partial), or to search under product/services you can enter a KEY WORD to receive a list of all records with that KEY WORD.

This list of categories is for suppliers under the general category of ANTENNA:

- Aluminum Tubing
- Antenna Dealer
- Antenna Manufacturer
- Antenna Part
- Antenna Relay
- Antenna Switch
- Antenna Tuner
- Balun
- Beam Heading Chart
- Coax
- Coax Sealant
- Diplexer
- Discone Antenna
- Dish Antenna
- Dummy Load
- Duplexer
- Fiberglass Material
- Gin Pole
- Ground Plane Antenna
- Guy Wire
- J Pole Antenna
- Ladder line
- Log Periodic Antenna
- Loop Antenna
- Mast
- Microwave Antenna
- Mobile Antenna
- Quad Antenna
- Rotator
- Rotator Control Cable
- Satellite Antenna
- Stacking Hardware
- Tower
- Tower Hardware
- Trap
- Tuner
- Twin Lead
- Twinlead
- Waveguide
- Wire
- Yagi Antenna

The advantage of using this online resource is that it is kept up to date by ARRL Headquarters staff. Be sure to check out the other services available on *ARRLWeb*. These inlcude ARRL advertiser links at **www2.arrl.org/ads/adlinks.html/html**. There is also a members-only section that includes a *QST* Product Review archive and a search function for *QST* and *QEX*. If you have not visited the members-only section, you will be pleased with the exclusive information and services available there to ARRL members.

About the ARRL

The seed for Amateur Radio was planted in the 1890s, when Guglielmo Marconi began his experiments in wireless telegraphy. Soon he was joined by dozens, then hundreds, of others who were enthusiastic about sending and receiving messages through the air—some with a commercial interest, but others solely out of a love for this new communications medium. The United States government began licensing Amateur Radio operators in 1912.

By 1914, there were thousands of Amateur Radio operators—hams—in the United States. Hiram Percy Maxim, a leading Hartford, Connecticut inventor and industrialist, saw the need for an organization to band together this fledgling group of radio experimenters. In May 1914 he founded the American Radio Relay League (ARRL) to meet that need.

Today ARRL, with approximately 170,000 members, is the largest organization of radio amateurs in the United States. The ARRL is a not-for-profit organization that:
- promotes interest in Amateur Radio communications and experimentation
- represents US radio amateurs in legislative matters, and
- maintains fraternalism and a high standard of conduct among Amateur Radio operators.

At ARRL headquarters in the Hartford suburb of Newington, the staff helps serve the needs of members. ARRL is also International Secretariat for the International Amateur Radio Union, which is made up of similar societies in 150 countries around the world.

ARRL publishes the monthly journal *QST*, as well as newsletters and many publications covering all aspects of Amateur Radio. Its headquarters station, W1AW, transmits bulletins of interest to radio amateurs and Morse code practice sessions. The ARRL also coordinates an extensive field organization, which includes volunteers who provide technical information and other support services for radio amateurs as well as communications for public-service activities. In addition, ARRL represents US amateurs with the Federal Communications Commission and other government agencies in the US and abroad.

Membership in ARRL means much more than receiving *QST* each month. In addition to the services already described, ARRL offers membership services on a personal level, such as the ARRL Volunteer Examiner Coordinator Program and a QSL bureau.

Full ARRL membership (available only to licensed radio amateurs) gives you a voice in how the affairs of the organization are governed. ARRL policy is set by a Board of Directors (one from each of 15 Divisions). Each year, one-third of the ARRL Board of Directors stands for election by the full members they represent. The day-to-day operation of ARRL HQ is managed by an Executive Vice President and his staff.

No matter what aspect of Amateur Radio attracts you, ARRL membership is relevant and important. There would be no Amateur Radio as we know it today were it not for the ARRL. We would be happy to welcome you as a member! (An Amateur Radio license is not required for Associate Membership.) For more information about ARRL and answers to any questions you may have about Amateur Radio, write or call:

ARRL—The national association for Amateur Radio
225 Main Street
Newington CT 06111-1494
Voice: 860-594-0200
Fax: 860-594-0259
E-mail: **hq@arrl.org**
Internet: **www.arrl.org/**

Prospective new amateurs call (toll-free):
800-32-NEW HAM (800-326-3942)
You can also contact us via e-mail at **newham@arrl.org**
or check out *ARRLWeb* at **www.arrl.org/**

FEEDBACK

Please use this form to give us your comments on this book and what you'd like to see in future editions, or e-mail us at **pubsfdbk@arrl.org** (publications feedback). If you use e-mail, please include your name, call, e-mail address and the book title, edition and printing in the body of your message. Also indicate whether or not you are an ARRL member.

On what other Amateur Radio subjects would you like the ARRL to publish books?

Where did you purchase this book?
☐ From ARRL directly   ☐ From an ARRL dealer

Is there a dealer who carries ARRL publications within:
☐ 5 miles   ☐ 15 miles   ☐ 30 miles   of your location?   ☐ Not sure.

License class:
☐ Novice   ☐ Technician   ☐ Technician Plus   ☐ General   ☐ Advanced   ☐ Amateur Extra

---

Name _____   ARRL member? ☐ Yes ☐ No
                                                           Call Sign _____

Daytime Phone (   ) _____   Age _____

Address _____

City, State/Province, ZIP/Postal Code _____

If licensed, how long? _____   e-mail address _____

Other hobbies _____

_____

Occupation _____

For ARRL use only	VHF/UHF
Edition	1 2 3 4 5 6 7 8 9 10 11 12
Printing	1 2 3 4 5 6 7 8 9 10 11 12

From _____
_____
_____

Please affix postage. Post Office will not deliver without postage.

EDITOR, VHF/UHF ANTENNA CLASSICS
AMERICAN RADIO RELAY LEAGUE
225 MAIN STREET
NEWINGTON CT 06111-1494

———————————————— please fold and tape ————————————————